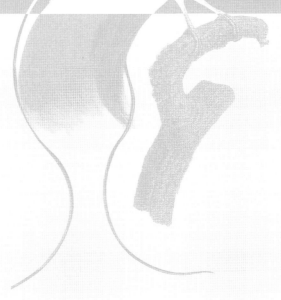

The History
of Ornithology

The History
of Ornithology

Valérie Chansigaud

NEW
HOLLAND

We do not completely understand a
science until we know its history.
Auguste Comte

To my mother,
friend to the birds.

An immense thank you to Agnès Hauck.

Front cover illustrations: Blue-and-yellow Macaw (*Ara ararauna*) by Edward Lear and (inset)
Red-breasted Merganser (*Mergus serrator*) by Louis Agassiz Fuertes
Spine illustration: Eagle Owl (*Bubo bubo*) by Edward Lear
Back cover illustration: Black Sicklebill (*Epimachus fastuosus*) by Richard Bowdler Sharpe
Illustration on the contents page: Noisy Pitta *(Pitta versicolor)*, from *Illustrations of Ornithology*
by Sir William Jardine

Published in 2009 by New Holland Publishers
London • Cape Town • Sydney • Auckland

www.newhollandpublishers.com

Garfield House, 86-88 Edgware Road, London W2 2EA, United Kingdom
80 McKenzie Street, Cape Town, 8001, South Africa
Unit 1, 66 Gibbes Street, Chatswood, NSW 2067, Australia
218 Lake Road, Northcote, Auckland, New Zealand

UK edition © 2009 New Holland Publishers (UK) Ltd
© Delachaux & Niestlé SA, Paris, 2007
Title of original edition: *Histoire de L'Ornithologie*
Published by Delachaux & Niestlé, Paris

10 9 8 7 6 5 4 3 2 1

A CIP catalogue record for this book is available from the British Library.

ISBN 978 1 84773 433 4

Publisher: Simon Papps
Translator: Joseph Muise
Editors: Beth Lucas and Marianne Taylor
Publishing Director: Rosemary Wilkinson

Printed by Mame, France

Contents

Introduction

In the animal world, birds are among the easiest organisms to observe: their size, colour and song all conspire to attract our attention. They have always been intimately tied to various aspects of human life: we hunt them and eat them, breed them and use their feathers for decoration; they can be found in religion, heraldry, symbolism, poetry, prophesies and fables; they are our everyday companions. However, it took a great deal of time and effort to better understand their behaviours (such as migration), to refine their classifications (even simply to separate them from bats), and to master their observation.

"Birds are the spoiled children of nature – the favourites of Creation ... Mankind has a profound sympathy with these little winged beings, which charm at once by the elegance of their form, the melody of their song, and the graceful impetuosity of their movements," wrote Louis Figuier (1819–1894) in 1868. It is true that the appeal of birds can easily explain the fact that the ornithological community is one of the best organized and largest in natural history. The knowledge obtained from the scientific study of birds has allowed for great progress in the biological sciences as a whole; in taxonomy, evolution, ethology, endocrinology, ecology, and other disciplines.

Unlike other scientific disciplines requiring heavy and expensive equipment, ornithology calls for light means of observation. This ease of observation, coupled with the charisma of its subjects, goes a long way towards explaining the large proportion of amateurs who make up contemporary ornithology.

The word ornithology (from the Greek *ornis* or *ornitho* for bird and *logos*, speech) meaning 'a treatise on birdsong', was adapted into English in 1655 from the Latin *ornithologia* that appeared from 1599. The first use of the term in its modern sense came in the 18th century following the publication of the works of Mathurin Jacques Brisson (the title page of his *Ornithologie*, 1760, is pictured here) and Buffon.

The history of ornithology is a human history

The history of a discipline cannot simply be a history of ideas, it is also and above all a history of those who devoted their lives to its development. To be a lover of nature and birds has never been, throughout history, an ordinary destiny. However, this attraction, which often manifests itself from childhood, must be supported and nurtured by the family circle and education, and must, come adulthood, find an outlet to express itself (through publications, through the putting together of collections, through participation in the work of museums and specialist associations and so on). It has always

With time, ornithological illustrations would be refined and gained greater realism. This became possible thanks to technical innovation (the move from woodcuts to the more precise copperplate techniques) and expanding scientific knowledge (above, an ostrich by Conrad Gessner from 1551, below, the same species by Nicolas Maréchal from 1801).

been an activity based around exchanges, as much of specimens as of knowledge. No one has ever amassed a fortune through the simple study of nature. To set one's sight towards natural history is also a sign of a certain courage and self-denial. Travelling naturalists have often risked and sometimes given their lives to the furthering of human knowledge.

Up until the middle of the 19th century, to train oneself in natural history meant undertaking medical study. It is therefore not surprising that we encounter numerous physicians amongst leading ornithologists. We find also a great number of servicemen and men of the church, as they are called upon to travel and hence to discover hitherto unknown fauna. Ornithology was for a long time a practice on the margins conducted largely in leisure time. In many cases it remains so today.

It is also a world of men. Women, who were unable to gain access to higher education until the very end of the 19th century, will be nearly completely absent from this history. Their role is not nonexistent, but is limited to being the assistants of their scientist husbands or illustrators (often very talented ones).

The history of ornithology is an international history

The compilation of a history of ornithology cannot limit itself to the borders of a single nation. Men, like ideas, travel. Let's look at a few examples, beginning with the development of natural history in Russia: our knowledge of the fauna of central Asia, the Urals and Siberia results largely from the efforts of naturalists of German origin, who found career opportunities in Russia unimaginable in their home country. The Frenchman Charles Lucien Bonaparte (1803–1857) contributed to the birth of American ornithology and Britain's John Gould (1804–1881) to Australia's, while the expeditions of Frenchman François Levaillant (1753–1824) to South Africa were financed in part by a Dutchman.

To recount the history of ornithology from a single nation's perspective would present a truncated and misleading account to readers. The conditions prove to be very different from one country to another: the British Empire provided great opportunities for exploration of fauna to many naturalist members of the colonial administration; the United States saw the establishment of many active associations, private and public; Germany benefited from a dense network of universities, some among the best in the world; and France became home to a mighty natural history museum that played a role in centralizing and organizing a number of explorations.

The history of ornithology is tied to the evolution of natural history

It is always a delicate matter to retrace the evolution of a discipline such as ornithology without relating it to the progress of the whole of natural history. For a long time, scientists with an interest in birds were equally passionate about other branches of natural history, including disciplines such as zoology, as well as botany and geology.

Little by little increased specialization would begin to emerge, even before the birth of the professions. This resulted largely from expeditions that led to ever increasing numbers of known species. Throughout the 19th century, the focus on heightened specialization was upheld by the establishing of institutions (such as museums and universities) to accompany new professions (curators and professors). The emergence of ornithology as a distinct discipline in its own right can be dated to the beginning of the 19th century. This would bring about the creation of learned societies uniquely devoted to the study of birds. An entire network would be built up, largely around ornithological journals and conferences.

The history of ornithology allows you to retrace the evolution of a discipline

Today, the word 'ornithologist' evokes images of an attentive and discreet protector of the environment sporting a pair of binoculars.

However, it has not always been this way. In its earlier days, every serious ornithologist was equipped with a good trigger; the study of birds was undertaken through the shooting of birds. A changing world, in the form of the uncontrolled industrialization of the 19th century, led to an awakening among ornithologists to the dangers facing numerous species and the imperative to protect them: the extinction of the Passenger Pigeon (*Ectopistes migratorius*) in the early 20th century was one of the factors that led to a greater awareness of the threats affecting birds. The methods of study changed and ornithologists would become true pioneers in species and habitat protection.

For a long time, the only lasting way to conserve the likeness of a bird was by depicting it visually, which accounts for the important role and proliferation of ornithological illustrations. Then, towards the end of the 18th century, methods were developed for preserving specimens by protecting them against the destructive effects of carrion-eating insects. This contributed to the growing interest in bird studies and the establishment of large public and private collections. With the collectors came a market for the collected object. Sold or exchanged, specimens frequently circulated amongst the early ornithological networks.

The work carried out by generations of more often than not amateur researchers is considerable. Today we know nearly all birds species that live on Earth, a fact that no other zoological group can boast (although the ease of their study and the relatively small number of bird species does make this easier).

To write a history of ornithology is to make choices

There is not a single prescribed method for writing history. A history of science, particularly when it is addressed to a wider readership like this one, must always err on the side of subjectivity and make arbitrary choices.

This work introduces men whose names are familiar to us through the common names of certain birds such as Temminck's Stint or Bonelli's Eagle. It also presents some lesser known aspects of ornithology such as the passion of the bird collectors of the 19th century, the role of illustrated works, the importance of museums and the first efforts to protect species, to understand migration and so on.

To finish, I would like to quote Sir Alexander Fleming (1881-1955): "A researcher's work is never complete. The mark of good work is when it opens the door to even better work, and quickly becomes eclipsed by it. The goal of research is advancement, not of the researcher, but of science."

Early ornithological illustrations represented birds as removed from their natural environment. Soon though, they would make way for spectacular illustrations that sought to represent birds in natural situations and in their environment (above, a Tawny Owl illustration by Xaviero Manetti from 1767-1776, below, an illustration of the same species by one of John Gould's artists from 1832-1837).

Antiquity
The first steps in the study of birds

Birds have always occupied an important place in human culture and society. Hunted for their meat, eggs and feathers, they are the subjects of intense observation as the many accounts testify. Omnipresent in the arts and religions, they are forever our companions. However, in spite of its precise nature, it took a great deal of time for this accumulated knowledge to be assembled into a coherent whole, and the scientific study of nature emerged rather slowly. Greek antiquity, nonetheless, left us with two monumental works: the first comes from the texts of Hippocrates on medicine, the second in those of Aristotle on physics and natural history. It is difficult to judge the quality of ornithological knowledge in antiquity outside of Aristotle's work, which consists of a mix of observations of nature, of symbolic and religious interpretations, of therapeutic concerns and philosophical musings. Despite being reproduced, compiled and commented on for centuries, the impact of these ancient texts on the development of Western thought would be in no way diminished.

Summary

The prescientific observation of birds

To hunt requires a close observation of nature: one must understand the ways and habits of animals to approach and trap them. However, the ancientness and accuracy of our knowledge of birds cannot be explained solely by the pursuit of the hunt.

Studies undertaken with certain tribes in New Guinea have revealed that they use 110 designations for birds while there are only 120 species in their immediate environment. Some are grouped under a common name, but others, rather difficult to distinguish in the wild, such as those belonging to the *Sericornis* genus, are well differentiated from one and other. Furthermore, male and female birds who display pronounced sexual dimorphism receive distinct names.

Knowledge of bird song is also passed on: we know that the song of divers (*Gavia* spp.) is imitated in the traditional chants of the Koyukon, a people of northern Alaska; a tribe in Costa Rica reproduces the song of the Riverside Wren (*Thryothorus semibadius*); Tuvan shamans in south central Siberia imitate the song of the Common Raven (*Corvus corax*) in their rituals.

Our fascination with the songs of birds was shared by numerous European composers including Vivaldi, Beethoven, Mozart, Brahms, Messiaen and many others.

Birds in mythology

Birds are not only the subjects of observation, they are also at the centre of an interest that blends art and religious beliefs. Their oldest depictions go back nearly 30,000 years (an owl carved into the cave at Chauvet in southern France). Also in France are the famous frescos of large animals such as bison, mammoths and Red Deer (*Cervus elaphus*) in the cave at Lascaux, along with drawings depicting many birds of prey and waterfowl. We also find small sculptures of birds carved from stone or bone. The bird motif can be observed in nearly all ancient cultures including the Sumerians, the Greeks and the Romans.

In ancient Egypt particular honours were accorded to certain birds. The sun god Ra (or Rê), creator of the universe, is often depicted as having the body of a man and the head of a falcon topped with a circle representing the sun, which is itself protected by a cobra. Another god, Horus, was depicted in the form of a man with a falcon head. Thot, the moon god, is

Classical representation of the Saker Falcon (*Falco cherrug*), guarantor of universal order and linked to the pharaonic monarchy, at the temple of Edfu in Horus's Egypt.

Licence Creative Commons/Hajor/Wikimedia Commons

The Yorck Project/Wikimedia Commons

Mural painting from the funeral room dating from the reign of Thutmose IV (1400 BC).

portrayed as an ibis (or as a baboon). More interesting from a modern ornithological perspective is the fact that ancient Egyptian frescos depict the Red-breasted Goose (*Branta ruficollis*) wintering in the Nile delta, which is not the case today. Images of the Greater White-fronted Goose (*Anser albifrons*) and the Bean Goose (*Anser fabalis*) can be clearly distinguished in the low reliefs of the famed tomb of Meïdoum.

Birds of prey occupy an important place in many cultures. We find eagle feathers amongst the religious objects of many American Indian peoples. Crows played a divinatory role for the Inuits; for the Haida and Athabascans, they participated in the creation of the world. More recently in North America, the dance of the crow or ghost dance appeared in 1890, following a vision by a Paiute. This dance sought to restore the traditional ways of life lost with the arrival of the Europeans.

As essential as the role played by birds is in these diverse mythologies, it is not with these peoples that we find the first manifestations of ornithology.

The founding role of Aristotle, the first ornithologist

It was only in ancient Greece that the systematic study of animals truly made its first appearance. As with other branches of natural history, **Aristotle** (384–322 BC) played a founding role.

A native of Macedonia, the young Aristotle entered into Plato's academy in Athens when he was 18 years old. With his bright mind, he began to teach soon thereafter. After Plato's death, he left Athens to return to Macedonia, where he became the tutor of Alexander the Great. A few years later, he returned to Athens where he created his own school, the Peripatetic School. The death of Alexander the Great drove him to leave Athens and he entrusted his school to Theophrastus.

If Aristotle included birds in his *History of Animals*, he did not seek to study them specifically. He sometimes cited inaccurate facts from earlier authors such as Aristophanes. However, he was interested in the experience of hunters, fisherman and farmers who reported details on the distribution, diet, song or wintering habits of birds, which he sorted into several groups based on the morphology of their extremities (feet clawed, webbed or with separated digits) and the nature of their feeding habits (insectivorous, granivorous, carnivorous).

In this way, Aristotle described 140 species of birds that observers in the Middle Ages would seek to identify, sometimes in vain. He seemed to not always know the appearance or the habits of the species he was describing and often ran into problems in dividing his groups (corresponding more or less to genera) into species.

In considering that anatomy is shaped by the requirements of the environment, he pondered the relationship between the size of animals and their climate and observed that seasonal changes in plumage can mislead one into believing in the existence of different species. He drew a parallel between the morphology of birds' wings and the paws of mammals. He stated that there is an analogy between the feathers of birds and the scales of reptiles. He was interested in the development of the embryo in eggs, but failed to mention whether the chick or the parent breaks the shell first.

Aristotle was also the first to look into the problem of the seasonal disappearance of birds, yet he wavered between two explanations: migration and hibernation. First, he noted that some species migrate great distances, such as Common Cranes (*Grus grus*), while others only migrate locally. He noted very rightly that individuals who migrate are fatter at the moment

Aristotle's bust. Marble, a Roman copy of an original Greek bronze by Lysippe (*c.*330 BC). Ancient Ludovisi collection.

Jastrow/Wikimedia Commons

14

of migration, which would only be confirmed and explained in the 20th century. He added that the reason for these seasonal mass movements is the sensitivity of these species to heat and cited the example of cranes and pelicans. Aristotle also mentioned altitudinal migrations, remarking that certain birds "descend into the plains during the winter and when it is cold, because they find the air to be more temperate; in summer, they retire to the heights of the mountains because the plains are blazing hot..."

He also remarked that a large number of birds that cannot migrate find refuge from the rigours of winter by hibernating. To support his theory, he cited groups of swallows found sleeping and stripped of their feathers in holes in the wall. Other species thought to adopt similar behaviour were storks, Woodpigeon (*Columba palumbus*) and Starling (*Sturnus vulgaris*). This theory would persist until the end of the 19th century and would only be contradicted through the use of ringing/banding.

Aristotle was convinced that certain species can transmute: he believed that the Black Redstart (*Phoenicurus ochruros*) changed its plumage come the arrival of autumn and became a Robin (*Erithacus rubecula*), and in so doing provided an explanation for the first named's seasonal disappearance.

He also echoed the tales of several legends. He wrote that if one punctures the eyes of a young Swallow, they heal as they grow and recover their sight. He asserted that nightjars (*Caprimulgus* spp.) suck the milk of sheep. He also got the incubation periods of certain birds of prey wrong (an error that would not be refuted for more than 2,000 years until the end of the 19th century). However, he refuted other legends by contradicting ancient authors who had asserted that vultures were born of the earth since their nests were not to be found. He revealed that, if we are unaware of their nests, it is because they are built in inaccessible locations.

Aristotle's zoology would not become a school of thought in his time and his work can be considered as an isolated occurrence in the ancient world. Even if he left no known disciples, his work would nonetheless play a fundamental role in the proliferation of modern European science.

Fictional portrait of Aristotle that appeared in the Nuremburg Chronicle.

Pliny the Elder, a great compiler

Pliny the Elder was born under Tiberius in 23 AD and died in 79, during the eruption of Vesuvius. The 37 books of his *Natural History* are not of the same quality as the work of Aristotle, being primarily a compilation of the period's

HISTOIRE NATURELLE
DE PLINE
TRADUITE EN FRANÇOIS,
AVEC LE TEXTE LATIN
rétabli d'après les meilleures leçons manuscrites,

De Notes critiques pour l'éclaircissement du texte,
& d'Observations sur les connoissances des Anciens,
comparées avec les découvertes des Modernes.

TOME PREMIER.

A PARIS,

M. DCC. LXXI.
AVEC APPROBATION, ET PRIVILEGE DU ROI.

One of the French translations of Pliny's *Natural History.*

Fictional portrait of Pliny that appeared in *Les vrais pourtraits et vies des hommes illustres grecz, latins et payens* (1584) by André Thevet (1516-1590).

Illustration of an eagle taken from a 9th century copy of *Physiologus* preserved in Berne.

knowledge of the animal world. They are the product of a credulous compiler rather than a critical mind. However, they are valuable because they bring together information from 2,000 works of all sorts, most of which have been lost.

The 10th book of *Natural History* is devoted to birds, beginning with the Ostrich (*Struthio camelus*), a bird Pliny claimed is closest to the class of quadrupeds. He also referred to the Arabian phoenix, the existence of which he doubted. The physical description that he provided is reminiscent of the Golden Pheasant (*Chrysolophus pictus*) according to Georges Cuvier. Pliny attested that cranes at rest during the night "place sentinels that hold a pebble in their claws; if the sentinel falls asleep, the pebble falls drawing attention to its negligence..." He asserted that storks would have the same habits. He took up Aristotle's avian hibernation thesis to explain their seasonal disappearance. He based his classification of birds on the form of their feet, but presented the details in a disorderly fashion.

In spite of the numerous legends that he reported, his work, widely read and reproduced at least until the 16th century, would serve as a standard text for generations of scholars.

Pliny is not the only compiler of this period. Aelianus (*c.*175-235) described 109 species of birds in *Perì zôión idiótêtos* (*Characteristics of Animals*), which brings together diverse anecdotes grouped in short chapters and without any classification.

A famous bestiary: Physiologos

Finally, it is necessary to mention a text that would profoundly affect medieval Christian thought: *Physiologos*, a 2nd century Greek bestiary by an unknown author from Alexandria. A Latin version, *Physiologus,* was compiled in 370 by Christian scholars. *Physiologos* describes animals and fantastical creatures (including stones and plants). Each description is an account accompanied by a symbolic interpretation. We find descriptions of real animals (like eagles, peafowl and ibises) and imaginary creatures (like the phoenix and sirens). *Physiologus* would be reproduced until the 15th century.

Antiquity would be followed by a long period in Europe where scientific thinking would fade into the background in the face of religious authority. It would take the rediscovery of ancient texts by Aristotle and Pliny, first commented on then criticized, to serve as the base for the birth and development of modern science in the West.

The Middle Ages
A long winter for science

In the Middle Ages the supernatural reigned supreme. Despite a passion for animals and plants that the arts and particularly the decorative arts bear witness to, their study progressed little over many long centuries. The evolution of European science can be summed up in several stages: the Middle Ages are, in comparison to antiquity, a period where science took a few steps backwards; over the 11th and 12th centuries Muslim science spread in the West and promoted the rediscovery of Greek authors; throughout the 13th and 14th centuries, scholastic science became established and spread before declining in the face of the emergence of an autonomous and modern science in the 15th century (as will be seen in the following chapter). The Middle Ages are, in essence, shaped by exterior influences, notably Arabic, but also, in other areas of science outside of ornithology, by Hebraic or Byzantine science. The history of bird study during this period can be followed via an analysis of Aristotelian texts, since translators were often commentators who enriched or disfigured the Greek work. One figure that stands out from this intellectual landscape is Frederick II of Hohenstaufen. He was head of the Holy Roman Empire, and a scholar whose observations were very advanced for the time.

Summary

The end of the ancient tradition

Bishop Braulio (left) and Isidore of Seville (right), illustration taken from a codex from the second half of the 10th century.

Isidore of Seville (*c.*560–636) succeeded his brother Leander at the episcopate of Seville, which was dominated by the Visigoths at the time. The period of peace that fell over the kingdom allowed him to transform the city into one of the most dynamic intellectual centres in Europe. His ambition was to undertake an etymological analysis of all words. His work, *Etymologiae*, was reprinted often until the Renaissance, a period in which his system, based solely on phonetic similarities, became the subject of critiques.

Isidore of Seville placed himself in the tradition of Pliny. His work perpetuated the ancient tradition as well as a certain number of its legends. Thus, he explained the softness of swans' vocalizations by the length of their necks and asserted that the saliva of cuckoos can produce grasshoppers.

From the ancient tradition to Muslim science

From 800 to 1100 most of Aristotle's works were translated and commented on in Arabic. Medieval Europe would only know Aristotle's work through these translations. It must be stressed that in the Arab world, zoology, like botany, has never been entirely independent from medical works or fantastical tales of nature.

Al-Jahiz (*c.*776–868 or 869), originally a fish merchant, began to frequent the mosque at Basra to study religion, but also philology, poetry and science. A prolific author, he wrote on an extremely varied range of topics. Only about 15 of these texts have survived, of which only one deals with animals. He attributed no real scientific significance to his grouping of animals into those that walk, fly, swim and crawl. However, he did question the paradoxes of his system: the grasshopper is not a bird though it can fly better than a chicken; the rooster is not a bird of prey since it lacks talons, however it hunts grasshoppers much in the same way that raptors hunt their prey. His goal was not to study nature through the eyes of a naturalist, but rather those of a philologist. His work would be referenced countless times and copied by subsequent authors.

A 15th century Iranian copy of one of Avicenna's *The Canon of Medicine*.

Avicenna (980–1037), born in the south of Persia, was known for his extraordinary memory, having memorized the entire Koran by his 14th year. He devoted himself to the study of medicine and science at a young age, but his patron died when he was 18 years old. He began to travel, and his great reputation slowly piqued the interest of a number of sovereigns. Avicenna participated in the Muslim intellectual movement that unravelled Greek culture. He translated the works of Hippocrates, Galen and particularly Aristotle into Arabic. His medical work was translated into Latin from the 11th century.

Another noteworthy translator of Aristotle's work is **Averroes** (1126–1198). This native of Cordoba became the physician of the Moroccan royal court. He undertook a translation of Aristotle's work into Arabic and added his own notes, whose quality merit the title of commentator.

Frederick II of Hohenstaufen, the learned Emperor

We know little of the early life and education of **Michael Scot** (*c*.1175–*c*.1232). We suppose that he is a native of Scotland where he taught Latin and ancient literature. Around 1217, he took a position as an astrologer and translator in Toledo, one of the most prestigious European cities in cultural terms. He first translated the Arabic work *Nur Ed-Din Al-Betrugi* (known under the Latin name of *Alpetragius*). His celebrity resulted from the translations into Latin of the works of Aristotle and Averroes's commentary. His translations would be referred to until the 15th century. Around 1220, he left for Italy where he would devote himself to medicine. He was offered several ecclesiastical offices in England and Ireland, which he refused, and around 1227 he joined to the court of Frederick II of Hohenstaufen.

Frederick II of Hohenstaufen (1194-1250), shown here with a bird of prey.

At the imperial court, he translated Avicenna's *Abbrevatio de animalibus* at the request of the Emperor and introduced him to the writings of Aristotle. It was at this time that Scot compiled his primary work, *Liber Introductorius*, a vast collection in which he covers astronomy, astrology, medicine, music and other subjects.

Frederick II of Hohenstaufen (1194–1250), christened *Stupor mundi* (the wonder of the world) by some, was king of Sicily from the age of 4 and Holy Roman Emperor from 1220 to 1250. This singular figure, passionate about the arts and sciences, began to surround himself with scholars, philosophers and Christian, Muslim and Jewish poets from the beginning of the 13th century and founded the University of Naples in 1229. He spoke nine languages and could read seven, at a time when most European monarchs were illiterate.

In spite of being excommunicated twice and treated as the Antichrist by Pope Gregory IX, he was the only European sovereign to enter Jerusalem without having spilled a single

Flight of the cranes, illustration taken from Frederick II's *De arte venandi cum avibus.*

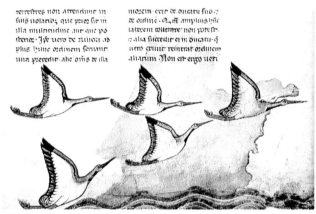

drop of blood; he was crowned the king of Jerusalem in 1229. He maintained good relations with his Muslim neighbours who considered him the most enlightened western sovereign, but his European contemporaries had a different point of view: Dante placed him in the sixth circle of his hell, reserved for heretics, despite the fact the Emperor's court was the origin of the blossoming of Italian literature. The end of his reign was marred by civil war in Italy and by his destitution under the actions of Pope Innocent IV.

Frederick II carried out numerous scientific experiments and was a fine observer of nature. He founded a veritable zoo in Sicily and raised giraffes (the first to arrive in Europe since antiquity), elephants, camels and leopards. Around 1240, he received an Umbrella Cockatoo (*Cacatua alba*) from the Moluccas from his friend the Sultan of Bagdad, but his true passion was the art of falconry. He was able to admire the Arab's mastery of the activity during his voyage to the Middle East. In his court, 50 falconers kept numerous birds of prey, some coming from faraway lands, such as the Gyr Falcons (*Falco rusticolus*) that he received from the north of Europe and Greenland.

Frederick II asserted the superiority of experiments to theories and consequently passed an edict in 1241 authorizing the dissection of cadavers, an edict that the religious authorities would hasten to repeal after his death. He gathered his ornithological observations in a book that he never completed, *De arte venandi cum avibus* (*On the art of hunting with birds*), the fruit of 30 years of observation. He did not hesitate to critique Aristotle and wrote: "We do not follow all the points of the prince of the philosophers as he had hardly, if ever, hunted with birds of prey; whereas we have always appreciated and practised this art … Aristotle speaks from hearsay, but the certainties cannot be separated from the gossip." Aristotle asserted that birds had no kidneys because urinating is not an independent function; Frederick II demonstrated the contrary through precise anatomical observations. He also refuted Aristotle's affirmations concerning the hibernation of certain birds (such as swallows or quails) in hollows.

In *De arte venandi cum avibus*, Frederick II not only covered birds of prey but addressed the breeding, mating and feeding habits of all sorts of birds. He experimented with artificial incubation and was particularly interested in the flight of birds, noting among other observations that the individual leading the V-shaped flight of cranes is frequently replaced by another.

The original manuscript of this remarkable work, which was ahead of its time, was burned by papal troops, but a copy was sent to Rome to be indexed. Historians attest that some of the 600 illustrations that feature in it are of the Emperor's own hand. More or less altered copies circulated from 1496, but it would not be until the end of the 18th century that the work would finally emerge from obscurity thanks to ornithologists Blasius Merrem (1761–1824) and Johann Gottlob Schneider (1750–1822). A facsimile was published in 1788 based on the

Falcons on their perches, with their hoods (above), falconers leaving for the hunt (below), illustrations extracted from Frederick II's *De arte venandi cum avibus.*

example preserved in the Vatican library. It was only in 1943 that a complete version was finally printed.

Dominican Albertus Magnus (*c.*1200-1280).

The blossoming of scholastic thinking

It was only in the beginning of the 18th century that Aristotle's texts were translated from Greek rather than Arabic manuscripts. **Albertus Magnus** (*c.*1200-1280) was the Dominican responsible for the translation of the zoological works of Aristotle, *De animalibus*, in 26 parts. He added a list of animals to his translation, largely inspired by the work of Pliny, although he did not hesitate to enrich it with his personal observations. It must be noted that Albertus Magnus was the first to refute the myth of the crustaceous origins of the Barnacle Goose (*Branta leucopsis*) and Brent Goose (*Branta bernicla*). These birds, which nest in the extreme northern regions of Europe, had never been seen mating or with eggs. The resemblance between the shape of the head of the Barnacle Goose and that of a crustacean, the Goose Barnacle which can be found attached to driftwood and the hulls of ships, perhaps explains the origins of this legend. Relating these birds to seafood had the added advantage of allowing their consumption during Lent. Albertus Magnus could refute this legend because he had seen geese breeding in the south of Germany. In spite of the observation and consumption of eggs by sailors on the Barents Sea near the end of the 16th century, Linnaeus nonetheless seemed to be giving credit to the legend by ascribing the name of *Anser bernicla* (goose issued from barnacles) to the Brent Goose and that of *Lepas anatifera* (who carries a duck) to the Goose Barnacle.

Amongst Albertus Magnus's contemporaries was his student **Thomas de Cantimpré** (1201–*c.*1272), whose *De naturis rerum* included accounts of 144 different birds (among them the bat). His work was translated into German by Conrad von Megenberg under the title *Buch der Natur* in 1475. During the same period, **Vincent de Beauvais** (*c.*1199–*c.*1265) orders Dominicans to write *Le Miroir de la Nature* (*The Mirror of Nature*), which traces the stages of Creation. The first animals it describes are domestic fowl. The text is taken from classical and contemporary authors, including Albertus Magnus.

This long period of The Middle Ages, more than a millennium, was marked by scientific stagnation. The work of Frederick II, as brilliant as it was, had little impact on ornithological thought. Modern science was truly born in the period that would follow, the Renaissance.

The Renaissance
The blossoming of science, expeditions and scientific publishing

The progress of ornithology during the Renaissance followed in the footsteps of natural history. The spread of printing throughout Europe would play an important role, enabling the publication of encyclopaedic reference works. The works of authors such as Gessner and Aldrovandi would draw on knowledge inherited from antiquity, while frequently adding original observations. Moreover, they established the necessity of illustration: authors and their illustrators sought to represent reality as faithfully as possible.

In their quest for exhaustiveness, encyclopaedists reported all the facts, real or imagined, including references to the harpy and the phoenix. Numerous aspects of the biology of birds remained a mystery. It was supposed that migratory birds which disappeared in winter hid in the waters of lakes or marshes or went to the moon.

Expeditions to ever more distant shores fed interest in the burgeoning science of zoology, contributing to a growing number of new species. The classification and description of these species led to an important realization: that the classical Greek and Roman authors would no longer suffice. Scientists would gradually gain independence from the grips of this classical heritage.

Ornithology was also to benefit from the contribution of the first true anatomists and the development of dissection. Their research would contribute to a gradual improvement in classifications and would finally force the separation of bats from birds.

Summary

The first printed illustration of birds dates from 1475 and was to be found in *Buch der Natur*, the work of Konrad of Megenberg (*c.*1309–1374). A hand-coloured edition would be produced three years later.

An edition of *Hortus sanitatis*, Strasbourg, *c.*1497. The work is attributed to Johann Wonnecke von Caub, better known as Johannes de Cuba, a physician based in Frankfurt am Main. The publication of *Hortus sanitatis* marked the beginning of a more systematic use of illustration in scientific works. Many illustrations drew inspiration from a medical treatise from the 12th century, the *Circa instans*.

The first printed works: the perpetuation of a tradition

It is often said that printing participated fully in the intellectual and scientific revolution that shaped Renaissance Europe. Nonetheless, the first editors and their readers long remained loyal to classical texts reprinted countless times by the likes of Albertus Magnus (see page 22) and Thomas of Cantimpré, or to compilations of the works of Aristotle, Pliny and others. Printing coexisted with the production of books by copyists for several decades. Illustrations were rare in the first printed books. Each of the 1,066 chapters of *Hortus sanitatis*, published in 1491, was preceded by one illustration.

These early illustrations were often of poor quality. An important first step in improving the quality of illustrations came with the opening of Otto Brunfels's (*c.*1488–1534) herbarium in 1530, which allowed the plants described to be identified. It would take more than 20 years for zoological works to attain a similar improvement.

The ancient works described animals that readers could not find in their surroundings. Moreover, these works had been translated and adapted a great many times. New works began to appear that provided names of animals in several languages, as is the case with the work of William Turner.

William Turner, author of the first Renaissance book on birds

We do not know the precise date of the birth of **William Turner**, son of a tanner, in a small town in Northumberland. It is likely to have been some time between 1510 and 1515. He graduated from Cambridge in 1529 and again in 1533. He experienced the Reformation under the dual influence of Hugh Latimer and Nicholas Ridley, who would both be burned in a public execution by Mary I in 1555. Turner went into exile in 1535 and became a physician in Italy in 1547. He followed the course of the great botanist Luca Ghini (1490–1566) and met Conrad Gessner (1516–1565) in Padova. The accession of Edward VI allowed him to return to England, where he became the physician of the Duke of Somerset and also received the deanship of Wells. In 1553, Mary I's succession to the throne forced him to return to exile. Following her death in 1558, he returned to England where he would pass away 10 years later.

Turner was responsible for about 30 works on the fauna and flora of Britain. They were very much a reaction to his contemporaries who contented themselves with translating the works of foreign natural histories, despite the fact that species described is these works were often absent from Great Britain. He is considered the father of British botany. During his first exile, he published *Avium praecipuarum, quarum apud Plinium et Aristotelem mentio est, brevis et succincta historia* in Cologne (1544). This was the first work completely devoted to birds and Turner's unique contribution to ornithology. Aside from a few glaring errors – he confused the Eurasian Bittern (*Botaurus stellaris*) with the Great White Pelican (*Pelecanus onocrotalus*) – certain parts of his text are highly accurate, much more so than those of his contemporaries. He did not restrict himself to describing species cited by Aristotle and Pliny, but included those that he had personally observed. He also asked his readers to bring other, less common birds, to his attention. It was William Turner's encouragement that led to the posthumous publication of *Dialogus de avibus et earum nominibus graecis, latinis et germanicis* by his friend Gisjbert van Langerack, also known as Longolius (1507-1543). However, this work is somewhat less valuable than Turner's, as Longolius contented himself with giving the names of birds in different languages.

Frontispiece of *Avium praecipuarum* by William Turner (1544). It is one of the first works to describe the birds of Europe. It is also the only ornithological work by its author.

Pierre Belon (*c.*1517–1564) is one of the central figures in the scientific revival: he travelled and observed peoples, fauna and flora. He based his books on fish and birds on his personal observations. Portraits of authors of scientific works were a new phenomenon that appeared around 1550.

Pierre Belon, scientist and traveller

One of the first scientific voyagers is the Frenchman **Pierre Belon** (*c.*1517–1564). We know that he was born in Souletière (near Mans), but the identity of his parents remains a mystery. Thanks to the protection of the Archbishop of Mans, René du Bellay (1500–1546), and then that of the Archbishop of Lyon, François II of Tournon, Belon was able to study at Wittenberg and Padova, where he studied under botanist Valerius Cordus (1515–1544).

In 1546, he set out upon a great expedition that would take him as far as the Middle East. Upon his return in 1549, he released an account of his travels under the title *Voyage au Levant, les observations de Pierre Belon du Mans, de plusieurs singularités et choses mémorables, trouvées en Grèce, Turquie, Judée, Égypte, Arabie et autres pays estranges*. In it, he described the fauna and flora of the countries he visited, but also the peoples that he encountered, the fortifications and ports, the knights and the foot soldiers. In recognition of his work, King Henry II rewarded him with a pension of 200 crowns and Charles IX provided him with lodgings in a castle situated in the Bois de Boulogne. It was in these woods that he would die, murdered at the hands of a prowler, at the age of 47.

In 1555, he released *Histoire de la nature des oyseaux, avec leurs descriptions et naïfs portraicts retirez du naturel*, which was of a higher quality than his previous work devoted to marine animals. He classified birds according to their behaviour and anatomy: birds of prey, aquatic birds, omnivorous birds, and small birds (which he subdivided into insectivores and granivores). This classification system is not terribly original,

The first recipe for the preservation of bird skins

"If a man curious of such things [birds], wanted to bring back the bodies from one country to another, this is the way to do it. The skin of the bird will have to be cut crosswise from the point of hard excrement, and all the internal organs removed, and covered with salt, and the inside of the stomach cavity stuffed, also the throat filled, then the bird taken by its feet. This will ensure that it will always remain whole with its feathers and shall not be consumed by worms, and if one sees that the salt does not dissolve, it will have to be wetted with a little strong vinegar ... Also, that it be a notice to all men reading this history and desirous of public good, that if they be in possession of some bird in their countries, that be not in this work, or of which is not discussed, preserve it according to what we have taught, and keep it to display in their cabinets, and if they were to be so inclined as to send it to us, we would be obliged."

Thus, Pierre Belon provides the oldest method of preserving the skins of birds (and he solicits the sending of travellers' specimens). He had the opportunity to put his method into practice during his voyage to Greece and the Middle East. Likewise, it is thanks to similar methods that he was able to study Brazilian Tanager (*Ramphocelus bresilius*) and Yellow-rumped Cacique (*Cacicus cela*) specimens brought back from Brazil.

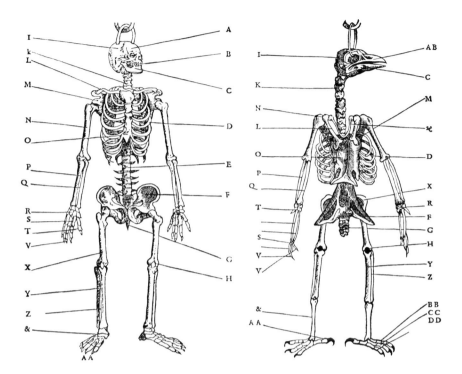

following the Aristotelian method. Despite knowing fewer languages than Conrad Gessner (he cited the names of birds only in Latin, Greek and French), his observations were much better than Gessner's (though Belon sometimes mistakenly described the Peregrine Falcon (*Falco peregrinus*) under different names based on its age, sex or distribution).

His text was supported by his own observations and it suggests that he practised a number of dissections. His famous illustrations, depicting the skeleton of a human and a bird side by side, demonstrate his great observational skills. He pointed out the similarities in bone structure, particularly in the wings of the bird. However, Belon had no knowledge of evolution. This comparison did not lead him to come up with even the most basic hypothesis to describe what he saw, and like other naturalists of his time, his analysis ended at the juxtaposition of two morphologies.

Belon was very diligent when it came to the quality of his illustrations and conscious of their essential role. The artwork in his books shows considerable progress on those that preceded them. He wrote in the introduction to his book: "As with the naïve drawings of snakes, fish, and birds that we reprint here: the nature that none other had seen is revealed to us. For as well as writing's ability to please the mind, and aid memory,

These illustrations, taken from Pierre Belon's *Histoire de la nature des oyseaux* (1555), have often been reused and analysed. They were the result of a curious mind interested in the similarities in bone structures between two organisms, not drawing any other conclusions. It would only be towards the end of the 18th century that comparative anatomy would be established. Pierre Belon is not, as he is often regarded, the father of comparative anatomy.

nous tenons quelque petit chien pour cópagnié, que faifons coucher fur les pieds de noftre liét pour plaifir : iceluy y auoit telles fois quelque Lion, Once, ou autre telle fiere befte, qui fe faifoyent chere comme quelque animal priué es maifons des paifants. Lon dit communement, que le Heron eft viande Royale. Parquoy la nobleffe Françoyfe fait grand cas de les manger, mais encor plus des Heronneaux: toutesfois les eftrangers ne les ont en fi grande recommandation. Il font

Pellos & Herodios en Grec, Pella & Ardea en Latin, Heron en Francois.

O' τὸ ον τοu δὴου οωφσνς φίλε γαλκεπτως ἰυντίζη ἡ ἡλύκ.πλζέι τι γὸ ἢ αὶ εια, ἱν οαειὸ ἀφιυτν, ἐν Τ ἰοδυλ μεὶν ἐχολαι, ἡ τλαλέ φιλλακε, ἡ ἰδιωγφὲε.τῶ κοφοιν ῆ, ὅιλοε παχχαιὶ ῆ τοὶς ἐα δι·τ εωετε, αἰςτῶ αἰρωλζε ῆ αυτι. ἑ ἀλωτινα, ιδιωγφε ῆ αυτιε τ νιντιε. καὶ κερφδω τὸ γαρ αὶὰ αὐτὰ κλάττοι, ἰωμιρς αιεε δ' ἴοι καὶ δωττογφρος και ἱωγρες τλυ φφυτν ηὺ ναι ἰγοι θιαιλυ ἦ τλαι κοιιλαιι αἰεὶ ὑφαηι. Ariſt.lib.9.cap.1.& 18.

fans cóparaifon plus delicats que les Grues. Il apert par le vol qu'on dreffe maintenant pour le Heron auec les oyfeaux de proyé, que les anciens n'auoyent l'art de fauconnerie fi à main comme on l'à maintenant. Ariftote a bien dit, au premier chapitre, du neufiefme liure, que l'Aigle affault le Heron, & qu'il meurt en fe deffendant. Le Heron fe fentant affailly, effaye à fe gaigner en volant contremont, & non pas au loing en fuyant, comme quelques autres oyfeaux de nuié re: & luy fe fentant preffé, met fon bec contremont par deffous l'ælle, fachant que les oyfeaux l'affomment de coups, dont aduient bien fouuét qu'il en meurt plufieurs

Combat du Heron auec l'Aigle.

imploring the faults of speech, and bringing certainty to doubtful matters: so can demonstration through figures and painting of written matter please the eye of the absent thing, almost as if it were present: illustrations bring the form and ways of things before our eyes ... there is no description or portrait of a bird in this work, that does not exist in nature, and that was not before the eyes of the painters."

Several French, Flemish, Italian and British artists were involved in the creation of these illustrations which would be often reused in the two centuries that followed. Pierre Belon only cites Pierre Godet, the painter responsible for a number of portraits of the French court, as the source of his illustrations. He had already signed the illustrations from the account of

Belon's travels, *Voyage au Levant*. The only illustration of a bird that figures in this work is a depiction of a European Bee-eater (*Merops apiaster*).

Conrad Gessner, the encyclopaedic work of a scholar

Conrad Gessner (or Gesner) (1516–1565) was born and died in Zurich. He was born of modest origins and became an orphan at an early age. His tutor, Oswald Myconius (1488–1552), allowed him to study at Zurich under Heinrich Bullinger (1504–1575). He completed his studies at Basel, Strasbourg, and Paris, and spent two years in Montpellier where he studied under Guillaume Rondelet (1507–1566), author of important works on fish.

Conrad Gessner (1516–1565) is one of the most famous of the 16th century encyclopaedists.

As was the custom of scholars of his time, he maintained regular correspondence with other naturalists, and exchanged specimens and information with them. He was known to have received fish jaws from William Turner (see page 25) and a nightjar skeleton from Pierre Belon.

He completed a degree in medicine in 1541 at Basel and began to assemble a cabinet of curiosities. In 1557, he was named professor of natural philosophy at Zurich and undertook a number of expeditions to study the flora, fauna and geology of the local area. Considered by many as one of the founders of mountaineering, he was well known for his excursions into the Alps. He died, perhaps of the plague, while treating compatriots struck by an epidemic.

Conrad Gessner is known for his extensive work as a bibliographer and encyclopaedist. He notably attempted to draw up a list of the works of all known authors and is considered one of the founders of philological studies.

However, it was his work on natural history, *Historiae animalium*, that he is perhaps most well known for. Gessner followed strict alphabetical order over 3,500 pages of the work and all species accounts are presented following the same logical structure. First, he gives the name of the species in different languages, and then describes its habitat and origin, provides some anatomical descriptions, reflects on the qualities of its soul, its medical uses, its value as food and ends with its use by philosophers and poets.

Naturally, he cited the works of Belon and Rondelet, but in fragmenting the original texts to follow his alphabetical arrangement, he lost the thread of reasoning that runs through the original authors' work and particularly the connections they make between related species.

Frontispiece of *Historiae animalium*, book III, by Conrad Gessner.

Plate depicting a grebe from the third volume of *Historiae animalium* by Conrad Gessner (1555). There are three uncoloured versions of the work in existence. Here, the colour was added by hand, page by page, a technique that would remain largely unchanged until the beginning of the 19th century. However, coloured books during this period were a rarity. The artist responsible for most of the illustrations in Conrad Gessner's work is the Strasbourg-born painter Lukas Schan.

Gessner proposed a rudimentary nomenclature and names the animals using two words, the second being a qualifier of the first, a system that would be taken up by other naturalists such as Rudolf Jakob Camerarius (1665–1721) and particularly Carl Linnaeus (1707–1778).

The section of *Historiae animalium* that deals with birds was released in 1555. Gessner described 217 species, each illustrated by a woodcut. Many of the illustrations were drawn from stuffed or mummified birds. What's more, we find depictions of some fantastical creatures like the griffin. The illustrations are far superior to those of Pierre Belon's publication that was released in the same year.

Gessner's work was critiqued, notably by Sir William Jardine (1800–1874), who considered it a literary and not a scientific work. However, Gessner remains one of the first to have assembled the component parts of natural history to study them systematically. His *Historiae animalium* was a veritable bestseller and would be published under a great many editions and translations until the 18th century.

Ulisse Aldrovandi, botanist, zoologist and teacher

Ulisse Aldrovandi (1522–1605) was born of a noble family of Bologna whose fortune allowed him not only to undertake numerous expeditions, but also to personally finance artists to illustrate his works.

After the early death of his father, Aldrovandi was initially expected to work in business. He became an apprentice at 12 years old, but he quickly realized that this path was of little interest to him. He travelled briefly to Spain before returning to study in Italy, at Bologna and Padova, which were the principal seats of scientific activity in Europe.

Having completed his diploma in 1553, he began to teach philosophy at the University of Bologna the following year. In 1560, he received the botany chair, only to widen the scope of the assignment to include the whole of natural history. It was around this time that he began to envision a vast project of books on this topic. During the summer and autumn, he scoured the neighbouring regions, often with his students, in order to observe the fauna and flora and to enrich his cabinet of curiosities.

In 1568, he founded the botanic gardens of Bologna. His courses, some of which were open to the public, brought him great renown. He later become an inspector of medicines and waged war against the bad preparations of many apothecaries, with the support of Pope Gregory XIII (1502–1585). In 1574, Aldrovandi released a treatise on medications.

One of the greatest achievements of his career was *Ornithologiae*, released in three volumes in 1599, 1600 and 1603. In it, he provided the names of birds in Greek, Hebrew, Arabic and Italian. He described their habits, habitat, explained how to capture and preserve them, how to cook them, their medical uses and, finally, explained their place in literature and mythology. The outline he followed echoed that of his near contemporary Conrad Gessner. A fourth volume, devoted to insects, would appear shortly after his death. He left large quantities of notes that were used by others taking up his work. They did not hesitate at times to make considerable alterations to the original text.

Ulisse Aldrovandi (1522–1605) played a significant role as a pedagogue and author. His works would be used for more than two centuries.

Frontispiece of *Ornithologiae* by Ulisse Aldrovandi, of which the first volume would be published shortly after his death.

Aldrovandi was very concerned with the quality of illustrations, as can be seen in this imprint of an Asian blue magpie.

786 · Vlyſsis Aldrouandi Pica Indica Mas.

Buffon was a staunch critic of Aldrovandi, whom he accused of hotchpotch writing. If the Italian author was wrong to seek to be exhaustive and to include everything that related to the animals that he described, he nevertheless made some good observations. He ascribed a great deal of importance to the quality of illustrations as this skeleton of a swan demonstrates.

Aldrovandi carried out high quality observations and, often, very original ones. He discussed the incubation period of eggs and provided incredibly precise anatomical descriptions, such as his description of the muscles that control beak movements in parrots. In comparison to Gessner's illustrations, they are of higher quality, but were often subsequently very poorly reproduced. Often drawn from dead animals, they seem rather artificial to modern eyes.

Aldrovandi's work was divided into 20 parts: 1. General remarks on eagles, 2. Description of different eagles, 3. General remarks on vultures, 4. General remarks on birds of prey, 5. Descriptions of different birds of prey (to which Aldrovandi adds the cuckoo), 6. General remarks on falcons, 7. Descriptions of different falcons, 8. On nocturnal birds of prey, 9. On birds who are a blend of birds and quadrupeds (like the ostrich and the bat), 10. On fantastical birds, 11. On parrots, 12. General remarks on crows and birds equipped with powerful beaks

The growth of cabinets of curiosities

The first great collectors were powerful men with sufficient resources at their disposal to pay for full-time illustrators. In this era, the preservation of specimens (particularly of birds) was not guaranteed; painting was the only way to ensure safekeeping of the memory of their form. Such was the thinking of Emperor Rudolph II (1552-1612), a keen student of the arts and sciences, who kept a rich cabinet of curiosities as well as a menagerie. Two Dutch painters, Joris Hoefnagel (1542-c.1600) and his son Jacob (1575-c.1630), were responsible for creating the canvases to immortalize the natural riches that he had acquired. Their paintings, 90 of which depict birds, are to this day preserved in Vienna. Scientific interest in these collections only began in 1868 when Georg Ritter von Frauenfeld (1805-1873) discovered a painting depicting a dodo (*Raphus cucullatus*) and other lost species (among them the only depiction of the Red Rail, *Aphanapteryx bonasia*).

Learned and wealthy Europe was taken by a passion for objects of natural history, but these were exhibited without logical order, unless it be an artistic one. They sought to display the most striking objects; often a bird would appear alongside shellfish, ancient shells, ethnological objects from America or Asia, a collection of antique items, or – the most highly sought – a teratological specimen.

It was not until the 18th century that collections would begin to be organized along more scientific lines, that they would slowly become more specialized, and that abnormality would be replaced by exhaustiveness. It was in this way that museums gradually evolved.

(such as toucans, birds of paradise, etc.), 13. On wild game birds, 14. On domestic game birds, 15. On birds similar to the preceding birds but who seek out water (like pigeons and certain passerines), 16. On blackbirds and related species, 17. On insectivorous birds, 18. On songbirds, 19. On web-footed birds and 20. On birds that frequent ponds and marshes (like herons, flamingoes, etc.).

His work, like Gessner's, would attract virulent critics. Georges Cuvier (1769–1832) claimed that "this work should be read with great care" and that he "does not seem to have the same critical mind" as Gessner. It is undeniable that Aldrovandi throws in descriptions of imaginary creatures or monsters with those of actual birds.

Due to their encyclopaedic nature, Gessner and Aldrovandi's works, in spite of the criticism they were subject to, would contribute greatly to the burgeoning scientific study of animals. These works allowed one to gain a general understanding of accumulated knowledge that future scientists could work with and improve on. This was a particularly important development as Europe was beginning to receive great numbers of specimens from newly discovered continents, which hinted at the existence of a fauna very different to that known in Europe up until then. This would give rise to a need to improve classifications and to perfect techniques for preserving and illustrating specimens. An entire literature would arise to meet these demands. In a few decades, the nature of observations would evolve and their quality would be considerably improved.

The first anatomical work: the important role of Italy

By the end of the 15th century, anatomy was an important subject of teaching. Here, the auditorium of the University of Padova, where Girolamo Fabrizio taught.

The Renaissance would bring about major changes in the study of birds, changes that would not have been as momentous without the equally formidable rise of anatomical science. One of the first figures to work on birds was **Girolamo Fabrizio** (1537–1619), better known as **Fabricius**, who was the first scholar since Aristotle to take an interest in the embryonic development of vertebrates and particularly of birds. Credited with the discovery of venous valves, he taught anatomy, occupying the renowned chair of the university of Padova. He counted William Harvey (1578–1657), the first to provide a scientific explanation of blood circulation, amongst his students.

It is the work of another Italian, **Marco Aurelio Severino** (1580–1656) that marks, for some, the beginning of comparative anatomy. In *Zootomia Democritaea* (1645), he

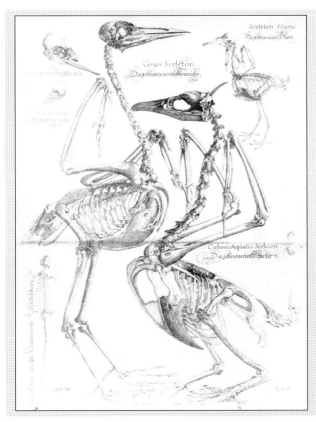

Plate showing the skeletons of a crane, starling, woodpecker, Wryneck (*Jynx torquilla*), cormorant and lizard, taken from *Lectiones Gabrielis Fallopii* (1575) by Volcher Coiter (1534–1576).

compares a diverse range of animals, including birds, discussed in the part titled *Ornithographie*. Severino was not searching for a universal method that would allow for the comparative study of the anatomy of organisms, but rather sought first to demonstrate the uniqueness of Creation. Aside from some very accurate observations, for example on the liver of geese, Severino commits a number of errors.

We know very little of the life of **Volcher Coiter** (1534–1576). A native of Groningen, he undoubtedly studied on the course offered by Leonhart Fuchs (1501–1566), the great botanist who taught at Tübingen. In 1559, he made his way to Bologna to enrol in courses offered by Aldrovandi. Converted by the Reformation, he was arrested by the Inquisition in 1566. After a year of detention, he left for Germany. From 1566 to 1569 he was the physician of the margrave of the Palatinate of Amberg and from 1569 to 1576, practised in Nuremberg. His most celebrated work is *Externarum et internarum principalium humani corporis...*, published in 1573 (in which he is the first to describe the female genital organs, and the dissection of human

Volcher Coiter (1534–1576), who, like other scientists of his time was persecuted for his thoughts on religion, was the author of a major work on avian anatomy.

cadavers that had been authorized by the Church in 1560). In *De avium sceletis et præcipius musculis* (1575), he is the first to propose a classification of birds based on anatomy rather than function. His observations on avian egg development appeared under the title of *De differentiis avium* (1575), in which he described the day-by-day development of the chicken embryo. This work would lead to his being recognized as one of the fathers of embryology.

Coiter's observations were very precise and he demonstrates depth of knowledge of the habits of birds. He established a connection between the beak shape and diet.

The advances in anatomy were to have a profound influence on classifications. After having contested little by little the existence of fantastical creatures, zoologists would establish a classification of living organisms based on anatomical criteria, as John Ray (see page 44) would in *De differentiis avium* (1690).

36

During the Renaissance, exotic birds became more and more numerous in artworks. Here is an example of Saint Jerome's chambers painted by Lucas Cranach the Elder (1526).

Exotic birds: source of inspiration and collectable objects

This painting by Roelant Savery (1576–1639) shows a large number of exotic species, including Ostriches and different parrots.

As we have seen with Pierre Belon, the techniques of preservation of bird skins were known and frequently used. However, they could not yet ensure long-term conservation, insects quickly reducing the specimens to dust. Drawing and painting remained the only way to durably preserve the memory of a species of bird.

Live exotic birds brought back by explorers piqued the interest of artists. A canvas from 1526 by Lucas Cranach the Elder (1472–1553) depicts the cardinal-archbishop of Brandenburg as Saint Jerome in his chambers with a lion and a parrot (no doubt a African Grey Parrot, *Psittacus erithacus*). Though this species was long known, the presence of a Yellow-crowned Amazon (*Amazona ochrocephala*) in *Breviarium Grimani*, completed in Flanders around 1500, suggests the rapid spread of foreign species. The presence of exotic birds in artwork would multiply, with African and American species featuring in the canvases of Jan Brueghel the Elder, (1568–1625), and Roelant Savery (1576–1639), particularly in scenes of nature and representations of earthly paradise.

It is also important to note the presence of now extinct species, such as the dodo, of which many specimens reached Europe as their multiple representations bear witness. The 16th century saw increased numbers of large and luxurious aviaries where rare, often exotic species were kept in captivity. Up until the 18th century, birds occupied a prominent position in menageries. Thus, the grand dukes of Tuscany, Francesco I de Medici (1541–1587), then Ferdinando I de Medici (1549–1609), maintained famous aviaries. Montaigne (1533–1592), who visited them in 1580, admired their weaver birds (*Euplectes* spp.) brought from the coasts of Africa. Ulisse Aldrovandi described many species from this same aviary in his *Ornithologiae*, some of which were brought from the Americas like the Helmeted Curassow (*Pauxi pauxi*), Great Curassow (*Crax rubra*), or the Northern Cardinal (*Cardinalis cardinalis*), others coming from Africa such as the Village Indigobird (*Vidua chalybeata*), Pin-tailed Whydah (*Vidua macroura*), and the Paradise Whydah (*Vidua paradisaea*). Rudolf II of the House of Habsburg (1552–1612) did not hesitate to spend 20 thalers to obtain the first specimen of the Chattering Lory (*Lorius garrulous*), brought back from the Moluccas. Next to these species chosen for their beauty, turkeys from America, guinea fowl from Africa, and pheasants and peacocks from China were raised. These species typically found their way to a plate.

The princes' aviaries would be emulated widely, first in Italy, then elsewhere in Europe. Catherine de Medici's aviary was known to contain three Ostriches, one of which would be dissected by Ambroise Paré (*c*.1510–1590). These aviaries were not solely the preserve of nobility: some bourgeois households, such as the rich Fugger family in Germany or Jacob Plateau in Belgium, kept rare animals.

It was for them that **Giovanni Pietro Olina** (1587–1645) would release *Uccelliera, overo discorso della natura e proprietà di diversi uccelli* in 1622 in Rome. Drawing his information from Belon, Gessner, Aldrovandi and others, Olina provided practical information on the maintenance of birds in captivity. Keeping exotic birds was an important investment and it was advisable to keep them alive as long as possible.

The birth of tropical zoology: de l'Écluse and Hernández

In the first accounts of expeditions, there were numerous descriptions of birds. Specimens brought back by voyagers were exhibited, like the parrots shown by Christopher Columbus alongside the riches of the new continent not yet known as America. A market for exotic birds existed in Europe throughout the 16th century: it was fashionable to own a parrot or some other exotic species, as can be seen in the work of artists from this period. We also know that one of Conrad Gessner's

gavit, eamque valdè brevem: ex pictura quam ad avis ipsius magnitudinem Illustrissimus Comes vivis coloribus exprimendam curaverat, Illustrissima verò Domina Sabina ab Egmonda, defuncti vidua, mihi commodabat, procurante ornatissimo & honestissimo viro Antonio de Flory, ipsius Consiliario, illius exemplar, in angustam formam contractum, in lignea tabella exprimendum (adjectà ipsius avis gemellà plumà, atque ad ejus pedes ovo) curabam, ut Lectoris oculis proponerem.

Emeu avis historia.

Hæc porrò avis, dum surrecto capite incedebat, quatuor pedum & aliquot unciarum altitudinem excedebat: nam collum à summo capite ad dorsi initium, tredecim pæne uncias erat longum, ipsum corpus binos pedes latum, femora cum cruribus usque ad pedum inflexionem, decem & septem uncias longa: corporis autem ipsius longitudo à pectore ad orrhopygium, trium ferè pedum erat. Pennæ, seu veriùs plumæ, totum avis corpus cum infima colli parte dorso & pectori proxima, atque femoribus, tegentes, perpetuò erant geminæ ex eodem parvo brevique tubulo prodeuntes, & sibi invicem incumbentes, superna quidem paullò crassior, inferna verò tenuior & delicatior, illæque variæ longitudinis, ut in similis avis exuvio, quod ornatissimus vir Christianus Porretus Leydensis pharmacopœus diligentissimus habebat, observabam: quæ enim in infima colli parte, breviores erant: quæ in medio corpore & lateribus, longiores, sex videlicet aut septem unciarum: quæ autem in extremo corpore, circa orrhopygium (nam cauda carebat) novem uncias longæ, & reliquis duriores, quanquam omnes duræ & rigidæ, attamen non latæ, sed angustæ, & rarioribus lateralibus pilis ex adverso sitis præditæ, nigri coloris, quæ tamen circum femora, ad cineraceum ferè tendentis, nigro permanente nervo, ut

in reliquis: eam verò formam & situm habebant illæ plumæ, ut à procul aspicientibus, non plumis, sed villis dumtaxat rectum ejus avis corium existimari posset, quale ursinum est, & alis carere, tametsi alas haberet, sed sub plumis, latera tegentibus, latentes, & quatuor majoribus pennis nigris præditas, ut in exuvio observabam, tametsi illæ abruptæ essent, nec de illarum longitudine certi quidpiam pronunciare possim, caules tamen confracti satis crassi, duriúque & solidi erant, atque in extimam alæ partem altè penetrabant: alæ verò pars superior corpori contigua, pennas sive plumas vestitrices, illis similes quæ in pectore, adhuc retinebat: ad hoc autem datas ejusmodi alas censendum est, ut currentem adjuvent, quia hanc avem non esse volucrem, nec à terra tolli posse, arbitror: crura in ambitu quinque unciarum crassitudinem superabant, crebrisque corticibus ceu squamis latis tecta erant, præsertim supra pedis inflexionem: pedes habebat crassos, duros, tribus crassis digitis præditos, pronâ parte veluti squamis tectos, supinâ prorsus callosos, quorum medius, qui reliquis longior, tribus articulis constabat, interior uno, exterior binis: omnium ungues valdè magni, binas pæne uncias longi, crassi, duri, & cornei: ejus caput exiguum, pro avis magnitudine, & fere glabrum, coloris ex atro cærulei, cum colli supremâ parte, in quo apparebant rari pili nigri: oculi paullò supra rostri fissuram magni, ardentes & truces, leonis oculis pæne similes; quos cingebant nigri pili, quemadmodum & meatus illos aurium parvos & detectos, quos ponè oculos habebat: rostri prona pars quodammodo arcuata erat, paulloque supra mucronem binis foraminibus, narium usum præbentibus, prædita; à cujus medio, ad capitis usque verticem porrectum diadema assurgebat corneæ substantiæ, tres ferè uncias altum, coloris ex flavo fusci, quod cum plumarum defluvio cadere, &, quùm plumæ renascuntur, novum crescere intelligebam: supina rostri pars, ab ejus scissura ad extremum mucronem, quinas uncias longa: anterior colli pars, quatuor pæne unciis infra rostrum, bina veluti membranacea palearia, ceu barbulas, propendentia habebat.

The frontispiece (above) and a page (left) from *Exoticorum libri decem* (1605) by Charles de l'Écluse. This book was the first work solely devoted to exotic fauna.

THE RENAISSANCE

Charles de l'Écluse (1525–1609) was a student of Guillaume Rondelet (1507–1566) of Montpellier. He spent several years in Spain, Portugal and Austria. Religious intolerance brought him to Germany in 1587 then to the Netherlands in 1593.

contacts informed him of the presence of numerous types of parrots in England in 1564.

Residents of the Netherlands were able to describe the first species that arrived in the ports during this period. Such was the case of **Jules Charles de l'Écluse** (1526–1609) or Carolus Clusius, known above all for his botanical work. In *Exoticorum libri decem*, he described many exotic plant and animal species. It is in this work that the Southern Cassowary (*Casuarius casuarius*), Magellanic Penguin (*Spheniscus magellanicus*), Greater Bird of Paradise (*Paradisaea apoda*), Lesser Bird of Paradise (*Paradisaea minor*), King Bird of Paradise (*Cicinnurus* *regius*), Red-fan Parrot (*Deroptyus accipitrinus*), Chattering Lory (*Lorius garrulus*) and Scarlet Ibis (*Eudocimus ruber*) are described for the first time. He also describes certain species that have since been lost such as the Great Auk (*Pinguinus impennis*) and the *Aphanapteryx* of Madagascar. Some science historians consider Charles de l'Écluse the founder of tropical zoology.

Naturalists would soon no longer be content to await the arrival of specimens and took part in explorations of newly discovered continents themselves. Some undertook private expeditions, others were commissioned by the authorities. Such was the case of **Francisco Hernández** (1514–1578) who, after practising at the prestigious Guadalupe monastery, became the physician of the royal court of Spain in 1565. He was selected by Phillip II (1527–1598) to conduct a mission of scientific study in New Spain. He arrived at Veracruz in February 1572, the head of an expedition that included a geographer, painters, botanists and specialists in indigenous medicine. On returning to Spain in March 1574, he devoted himself to studying the collection that he had amassed. He was particularly interested in the knowledge of the indigenous peoples, as well as their language, Nahuatl, that he learned during his stay in Mexico. He would face many hurdles on the road to publishing his observations. After his death, the king entrusted their publication to an Italian editor, in what would be a failed attempt. The greater part of Hernández's original illustrations disappeared in a fire in the library of El Escorial in 1671.

Hernández described more than 230 species of birds, but the absence of illustrations renders their identification very difficult. Johann Georg Wagler (1800–1832) would be the first to attempt to identify them, without much success. However, the fact that Hernández provided the Nahuatl names for the birds sometimes allowed them to be identified. For two centuries, Hernández remained the only source of information on the fauna of Mexico, the Spanish imposing tight controls on information coming from Central America. The exploration of Mexico would only become possible in 1821, when the country gained independence from Spain. In 1823, William Bullock (1773–1849), a naturalist, brought back the first collection of Mexican birds to Europe. His descriptions would be taken up by Jonston, Willughby and Ray in the 17th century and Brisson, Buffon and Linnaeus in the 18th century.

Illustration of a bird of paradise taken from the 1651 edition of *Nova plantarum, animalium et mineralium mexicanorum historia* by Francesco Hernández.

The 17th century
The founding work of John Ray and Francis Willughby

D uring this century, zoology would not be subject to the amount of upheaval and progress that botany would experience. It was a period of transformation, marked by the discovery of a growing number of specimens, particularly exotic ones. The expeditions undertaken, whose principle goal was to establish trade routes and relationships, counted amongst their crew not only a cartographer, but also a naturalist. The taste for exoticism was not the sole preoccupation: interest in domestic fauna began to grow, and would lead to the publishing of works on European fauna. The scientific study of birds in the 17th century is largely dominated by two men: John Ray and Francis Willughby, who established the first modern classification. Among the scientific advancements realized during this century, the microscope allowed the discovery of structures that were once invisible to the naked eye, such as the structure of feathers. The theory of spontaneous generation began to be contradicted by the works of Francesco Redi (1626–1697), Marcello Malpighi (1628–1694) and Jan Swammerdam (1637–1680). It would be definitively rejected in the 19th century owing to the work of Louis Pasteur (1822–1895). The study of chicken eggs allowed important discoveries about embryonic development and the formation of the nervous and circulatory systems. The study of nature was no longer undertaken by men in isolation. Scientists gathered: the Accademia dei Lincei was founded in Rome in 1603, the German Academy of Sciences Leopoldina at Halle in 1652, the Royal Society in London in 1660, and the Académie des Sciences in Paris in 1666. These societies and academies would be followed by many others, often more regionally based, reflecting the growing popularity of natural history.

Summary

In his ornithological work, John Jonston (1602-1675) compiled the work of ancient authors, presenting no new ideas.

John Jonston:
the end of the encyclopaedists of the Renaissance

John or **Jan Jonston** (1603-1675) was a physician who split his time between England and his native Poland, which provided him with many opportunities to travel through Western Europe. He released a series of works on natural history, beginning in 1657. Volume six was devoted to birds. It was very quickly from Latin into German, English, Dutch and French. His last edition appeared in 1773.

His work compiled earlier authors, often from antiquity as well as the Renaissance, such as Pliny, Gessner and Aldrovandi.

Plate taken from Volume IV of *De avibus de l'Historiae naturalis* by John Jonston. The engravings were the work of the famous German engraver Merian.

He added very little in terms of new material. Following the example of his predecessors, he described fantastical creatures like the griffin next to a description of the hippopotamus. The most remarkable feature of his work was its illustrations. These were taken from the works of his predecessors and were signed by the German engraver Matthäus Merian the Younger (1621–1687), born of a prestigious line of artists and brother of Maria Sibylla Merian (1647–1717), the first woman entomologist.

Jonston brought no new observations to light, but his text was unique in that it only dealt with the scientific aspects of animal studies, abandoning the humanist digressions that had previously accompanied the works of the encyclopaedists.

Walter Charleton (1619–1707) borrowed many elements from the encyclopaedists of the previous century (such as the survey of previous works and the quest for exhaustiveness). He obtained his title as medical doctor at Oxford in 1631 and became physician of Charles I of England (1600–1649), then Charles II (1630–1685). He published works on extremely varied subjects (he also tried to demonstrate that Stonehenge was the work of the Danes, who used it to crown their kings). He released *Onomasticon Zoicon* in 1668, in which he attempted to classify all known animals in the first part. The second dealt with anatomy and the third with mineralogy.

Charleton studied the birds of Charles II's aviary, as well as the collections of the Royal Society. His classification, inspired by the Aristotelian model, divided birds into two main groups: terrestrial and aquatic.

The former are classed according to their diet: carnivores (like eagles, cuckoos, or emus); granivores subdivised into those who bathe in dust (like chickens or geese), those who bathe in dust and water (like pigeons or sparrows), those who sing (like larks or chaffinches); berry eaters (like blackbirds or starlings); and finally insectivores, which are divided into non-singers (like woodpeckers or nuthatches) and singers (like tits or nightingales). The aquatic birds are classed as palmipeds or fissipeds. The latter are in turn subdivided into fish eaters (herons or storks), insectivores (like rails or moorhens) and vegetarians (like cranes).

His ornithological writings would not become very successful, as they would very quickly be supplanted by Francis Willughby and John Ray's *Ornithologiae*.

Frontispiece of *Historiae naturalis de avibus* by John Jonston (1657). The illustration was signed by Kaspar and Matthäus Merian.

Title page of *Onomasticon Zoicon* (1668) by Walter Charleton.

Another look: the first works on regional fauna

At the beginning of the 17th century, another sign of the break with the authors of the Renaissance became evident with the appearance of studies on regional fauna. Thus, **Caspar Schwenckfeld** (1563–1609), born in Silesia, released the first work on regional fauna in 1603, *Theriotropheum Silesiae*, of which the forth volume is deals with birds. A physician, he worked initially in Greiffenberg (modern Gryfów Slaski), in Silesia, then at Hirschberg (modern Jelenia Góra). He was particularly interested in the habits of animals because they were thought to demonstrate divine wisdom. He drew largely from the works of Gessner, Aldrovandi and others, from which he faithfully reproduced all their assertions, even the most false ones. He described about 150 species rather precisely and, whenever possible, provided complementary information on ecology, song, number and colour of eggs, etc. Schwenckfeld devised a possible classification of birds based on five criteria: habitat (which he divides into seventeen ecological units); mobility (partial and total migratory species, sedentary species); feeding (he distinguishes nine types of diet: carnivore, frugivore, etc.); structure of the feet (between fissipeds and palmipeds); and colour. Nonetheless, he considers this system overly complicated and prefers alphabetical order, in the manner of Gessner.

The first study of British fauna was published in 1666 under the title *Pinax rerum naturalium Britannicarum*. Its author, **Christopher Merrett** or **Merret** (1614–1695), does not make any original observations, rather relying on the work of Aldrovandi and Jonston. Like them, he classed birds first into terrestrial and aquatic species. The former are divided into several subgroups according to their diet (carnivore, vegetarian, frugivore, insectivore). Aquatic birds were classed as palmipeds or fissipeds (which are further subdivided based on their diet). Merrett described 165 British birds, and is credited with having contributed to popularising the study of nature.

John Ray and Francis Willughby: the precursors of modern ornithology

These two men played an important role in the advancement of knowledge in natural history, as much in botany as in zoology. Their work turned the classification of birds based on Aristotle's work completely on its head.

John Ray (1627–1705) was born of modest family. His father was a blacksmith. Thanks to a system of bursaries granted by the vicar of Braintree, he was given the opportunity to continue his studies at Cambridge. After a year at St. Catherine's College, he transferred to Trinity College. He would remain there for 17 years, first as a student, then as a lecturer. He took holy orders in 1660. It is at this time that his interest in botany materialized and he released a work on the flora of Cambridgeshire and the surrounding area. He would be obliged to leave Cambridge in 1662, because of a royal decree that would distance many church figures from British universities.

John Ray (1627-1705).

TAB. XXXVII

Turdus simpliciter dictus. The Mavis or Song Thrush.

Turdus pilaris. The Fieldfare.

Merula. The Blackbird.

Passer solitarius mas Aldrov.

Merula torquata. The Ring Ouzell.

Sturnus. The Stare or Starling.

Francis Willughby (1635-1672).

Plate of European thrushes and starling taken from *Ornithologiae libri tres* by John Ray and Francis Willughby.

Frontispiece of *Ornithologiae libri tres* by John Ray and Francis Willughby. It was first published in Latin in 1676 before being translated into English two years later.

It was during his stay in Cambridge that Ray first befriended **Francis Willughby** (1635–1672), a friendship that would continue until Willughby's premature death. Ray and Willughby began to explore the coast of the British Isles in 1660, to undertake a careful observation of the mating habits of marine birds. Willughby then financed a trip through Europe from 1663 to 1666.

This expedition would allow them to study birds that were unknown to them in Great Britain. They were particularly interested in the market stalls of the cities they visited, where a great number of bird species hunted in the surrounding area were available for sale. This is how Ray and Willughby discovered the Hazel Grouse (*Tetrastes bonasia*), the Golden Oriole (*Oriolus oriolus*) and the Black Stork (*Ciconia nigra*) in Germany, the European Serin (*Serinus serinus*), Citril Finch (*Serinus citrinella*) and the Crested Lark (*Galerida cristata*) in Austria, the Water Rail (*Rallus aquaticus*), the Capercaillie (*Tetrao urogallus*) and the Great Bustard (*Otis tarda*) in Italy. These freshly killed specimens allowed them to study their appearance, but also their anatomy. Attentive observers that they were, they were interested in the birds' internal and external parasites, as well as the contents of their stomachs. They took advantage of their trip to acquire illustrations and works like *Vögel, Fisch, und Thierbuch* (1666) by Strasbourg native Leonhard (Linhardt) Baldner (1612–1694), one of the first works printed on fishing and hunting.

The death of Willughby's father in late 1665 obliged him to return to England. Ray would follow the next year. Back in England, the two men began to study the specimens they had collected and gathered their observations with the objective of releasing an extensive work of natural history. In 1671, Willughby began to prepare a trip to North America in order to complete their observations. It is at this point that his health began to deteriorate and he was forced to abandon the project. He died in July 1672. John Ray was informed of the news upon his return from a plant-collecting expedition.

Francis Willughby proved very generous towards his friend. As well as a comfortable pension, Ray was entrusted with the education of his late friend's children. He committed himself to the publication of their shared work from then onwards, and would spend four years compiling and completing the notes left by Willughby. He married in 1673 and, once relieved of his duty to educate Willughby's sons in 1675, he returned to his native Black Notley in Essex, where he would remain until his death.

Ray published *Ornithologiae libri tres* under his friend's name. It is impossible to know the exact contribution of each of the two scientists, Ray having been very modest. It is undeniable that Ray's experience in botanical classification contributed greatly to the drawing up of the system employed in zoology. Since the Renaissance, the progress achieved by botanists had allowed them to distance themselves from the classifications inherited from antiquity. Zoology dragged behind botany on this front, and Ray was the first to apply similar approaches to it.

This work is considered seminal to the development of modern ornithology and their classification would remain without equal for a long time. Even the classification used by Linnaeus, despite qualifying as natural, is very much inferior.

It is known that Ray completed Willughby's notes with his own observations, notably by virtue of the specimens that he received from his contacts. Thus, he was able to add the Black Tern (*Chlidonias niger*), Common Pochard (*Aythya ferina*),

47

17TH CENTURY

The ornithological work of John Ray and Francis Willughby would see numerous editions and adaptations until the beginning of the 19th century. Here, a plate by François-Nicolas Martinet, added to the French edition in 1767. Above, the European Nightjar (*Caprimulgus europaeus*), centre, Tawny Owl (*Strix aluco*), below, Eagle Owl (*Bubo bubo*).

When bats were birds

For a long time the term bird referred to all flying creatures (which included bats) in much the same way that the word fish signified numerous creatures living in the water, including dolphins and whales.

Thus, Walter Charleton (1619-1707) classes owls, shrikes, cuckoos, parrots, crows, ostriches, cassowaries and bats as carnivorous terrestrial birds. Nehemiah Grew (1641-1712) asserted that bats were intermediary creatures between mammals and birds. In 1657, John Jonston featured birds and bats on the same plate (right).

Advances in anatomy allowed the progressive improvement of classifications. During the same period, John Ray and Francis Willughby would exclude bats from their *Ornithologiae* (1676).

Smew (*Mergellus albellus*), Shellduck (*Tadorna tadorna*) an the Eurasian Wigeon (*Anas penelope*).

In *Ornithologiae*, birds are first classified as terrestrial and aquatic. The latter are subdivided into those that frequent wet areas and those that swim. These two groups are then subdivided into smaller groups. The terrestrial birds with hooked beaks and talons are subdivided into the carnivores (like the birds of prey, which are themselves separated into nocturnal and diurnal species) and the frugivores (like parrots). The groups developed by Ray and Willughby are often very similar to current classifications. This classification would be followed by a number of ornithologists up until Linnaeus (see page 78). In an attempt to better identify the different species of birds, the authors proposed a dichotomous system of determination very similar to modern approaches.

The quality of its illustrations, the work of many artists, was inferior to those in of Walter Charleton's (1619–1707) *Onomasticon Zoicon* (1668) (see page 42). The illustrators do not seem to have had a sound knowledge of the birds they were depicting. The illustrated species were often grouped onto a single plate without any indication of scale. The birds, drawn from dead specimens, do not seem to have natural bearings and the backgrounds are extremely simplified.

GEORGI MARCGRAVI

CAP. VI.

Iabiru. Iabiru guacu. Manucodiata.

IABIRV Brafilienfibus, Belgis vulgo *Negro.* Avis hæc magnitudine fuperat Cignum. Corpus illius quatuordecim digitos longum : collum totidem, & brachii humani habens craffitiem. Caput fatis magnum, oculi nigri, roftrum nigrum directe extenfum, & fuperius verfus extremitatem paulum incurvatum; undecim digitos longum, duos & femis latum, ver-

fus exteriora acuminatum; eftque fuperior roftri pars paulo altior & major inferiori. Caret lingua & fub gutture ingluviem habet mediocris magnitudinis. Crura longiffima, duos nimirum pedes : fuperiora enim unum pedem & digitum longa, & mediam partem pennis nuda; inferiora undecim digitos. Sunt autem crura recta, nigricantia, & quafi fquamata, digitum medium craffa. In pedibus digiti quatuor, tres anterius, unus pofterius verfus; quorum medius quatuor digitos longus, cæteri paulo breviores. Tota avis veftitur albis pennis inftar cigni aut anferis. Collum fere totum, nimirum octo digitorum, longitudine à capite numerando, caret pennis; ac hujus medietas cum capite tegitur nigra cute, reliqua alba cute. Sed puto in cute hæfiffe pinnulas albas, & fuiffe abreptas. Cauda lata definit cum extremitate alarum.

Page on the Wood Stork (*Mycteria americana*) taken from *Historia naturalis Brasiliae auspicio et beneficio illustratis* (second edition, 1658) by W. Piso and G. Markgraf.

Ray made use of accounts of Sir Hans Sloane's (1660–1753) travels in Jamaica (1687–1689) and Paul Hermann's (1646–1695) travels to Ceylon (1672–1680), and studied the collections of the Royal Society and those of Sloane.

Ray released several major works on botanical classification. He also continued his zoological research with the publication of *Synopsis methodica animalium quadrupedum et serpentini generis* (1693) and *Synopsis methodica avium et piscium* (1713). The latter was published posthumously in 1713 thanks to his friend the canon William Derham (1657–1735), who adds the first list of birds of India, *Avium Maderaspatanarum* (the birds of Madras) compiled by James Petiver (*c.*1663– *c.*1718).

A French translation, with considerable improvements by French physician François Salerne (*c.*1705–*c.*1760), was published in 1767 titled *L'histoire naturelle, éclaircie dans une de ses parties principales, l'ornithologie, qui traite des oiseaux de terre, de*

Monogram of the Vereenigde Oostindische Compagnie (VOC) or Dutch East India Company.

mer et de rivi re tant de nos climats que des pays étrangers. It was illustrated by François-Nicolas Martinet (*c.*1760–1800) who also illustrated the works of Brisson and Buffon.

The work of Ray and Willughby marks an important transition. It is the last attempt to gather all knowledge on birds into a single work. After them, authors would specialise, taking up anatomy, physiology or classification. From this period, the assembled observations had become so vast that it would be impossible for one author to achieve a complete knowledge of them. Even Ray and Willughby are not exhaustive in their discussions of anatomy.

The first scientific voyagers: Europe sets out to discover the world

In the 16th century, Francisco Hernández (see previous chapter) was one of the first to be appointed to oversee a scientific mission for the Spanish crown. Many others would follow in his footsteps. The expeditions of the 17th century more and more frequently involved scientific components, in an attempt to gather knowledge and collect specimens. The goal was clearly to discover species that could be of economic value. It represented a change in the willingness of European States to organize the exploration, exploitation and management of new territories. Such was the case with the English East India Company in 1600, quickly joined by the powerful Dutch company with its famous acronym, VOC. The French equivalent was not created until 1664 and would never enter into competition with the other companies. They were very active organizations and would finance or assist scientific missions.

It is in this environment that John Maurice of Nassau-Siegen (1604–1679), governor general of the Dutch colonies in Brazil, commissioned his personal physician **Willem Piso** (1611–1678) to organize a scientific mission seeking to understand the geography, flora and fauna of the new continent. Among the crew was **Georg Markgraf** (1610–1644) whose primary responsibility was geography, but he was also charged with natural history. He was accompanied by two artists, Frans Post (1612–1680) and Albert Eckhout (1610–1665). Markgraf would die of fever, most likely malaria, at 34 years of age, upon his arrival in Angola for another expedition.

The botanical and medical observations of the expedition would make their way into the 1648 work *Historia naturalis Brasiliae.* A few years later, Piso would take it up again; he

Frontispiece of *Historia naturalis Brasiliae* (1648) that brought together observations from the first scientific expedition to South America.

published an improved version in eight volumes titled *De Indiae utriusque re naturali et medica* (1658). Three volumes deal with botany, four with zoology, all drawn from Markgraf's notes. The eighth volume discusses the geography of the regions explored and their residents. Piso deliberately omitted Markgraf's name from the cover; worse, he denigrated him by presenting him as his servant, describing him as an alcoholic and accusing him of embezzlement. Moreover, the woodcuts carved from Markgraf's original illustrations are of mediocre quality, making the identification of species quite difficult. It would only be in 1786 that the quality of Markgraf's observations would be recognized for their just value. J. G. Schneider (1750–1822), who had already brought the work of Frederick II of Hohenstaufen to light a few years earlier (see page 19), discovered and drew attention to Markgraf's original paintings. These allowed the species observed to finally be identified and the reader to better appreciate the great observational abilities of the author. Markgraf's illustrations were studied by Johann K.W. Illiger (see page 140), H. Lichtenstein (see page 141) and finally by Adolf Schneider (1881–1946), who identified them in 1938.

Frontispiece of *De Indiae utriusque re naturali et medica* (1658), attributed solely to Willem Piso.

Markgraf had described 133 species of birds and the information that he assembled would remain the only details available on the avifauna of South America for several decades. They would be used by numerous scientists, including Ray, Brisson, Buffon, Linnaeus and Johann Friedrich Gmelin. The two latter would designate no fewer than 45 species based solely on Markgraf's descriptions. Cuvier considered the work to be "a classic book that one can consult with complete confidence in all that it contains."

The study of the natural history of the West Indies began with **Jean-Baptiste du Tertre** (1610–1687) who set out on missionary work in 1640. First a Dutch soldier in the marines (which gave him the opportunity to visit Greenland), he became a French infantryman during the siege of Maastricht in 1633. He returned definitively to France in 1656 and penned several works on the West Indies and their inhabitants. He described certain birds with much lyricism, such as "the admirable hummingbird, admirable for its beauty, for its smallness, its nice smell, and its way of life. That this bird be wonderful for its size, and its plumage; it is no less worthy of admiration for the activity of its flight, which is so fast and so precipitous, that by proportion the largest birds do not break the air with as much force and do not make such a resonant noise, as that which excites this amiable little hummingbird in the beating of its wings. ... It lives only from the dew, which it sucks from the flowers of trees with its tongue, which

Frontispiece of *Histoire naturelle et morale des îles Antilles de l'Amérique* (1658), one of Jean-Baptiste du Tertre's (1610–1687) works on the West Indies.

The disappearing specimens

By the 17th century, taxidermy had spread widely, but the preservation of stuffed birds had not been fully mastered. The specimens represented in the catalogues of cabinets of curiosities, like that of Ole Worm, have not survived, destroyed by insects like the Larder Beetle (*Dermestes lardarius*) and Museum Beetle (*Anthrenus museorum*). Worm's collection, famed throughout Europe, contained a number of rarities like the Great Auk (*Pinguinus impennis*). A similar fate would befall the stuffed dodo from John Tradescant the Elder's (*c.*1570-1638) splendid collection, that would be further improved by his son, John Tradescant the Younger (1608-1662). This dodo specimen, exhibited at Oxford for a century, disintegrated into a pile of dust over the course of the 18th century.

It is owing to these difficulties that some species like the dodo, now extinct, are only known through their pictorial representations. However, illustrations were often copied from one book to another and sometimes altered. Above right, an illustration taken from *Exoticorum libri decem* (1605) by Charles de l'Écluse (see page 38), who no doubt had the opportunity to observe one of the live dodos brought back from the Indian Ocean. Below right, the same bird depicted in *Ornithologiae* (1676) by Ray and Willughby. The body is considerably podgier, the posterior has gained feathers and the beak is transformed.

Dodo.

is much longer than the beak, which is hollow like a straw, and the size of a minute needle". The second volume of his *Histoire générale des Antilles habitées par les Français* (1667-1671) is entirely devoted to natural history. Du Tertre makes use of extracts from the work of Piso and his writings were also used by other authors. He had no scientific training and his observations, while pioneering, were not of the best quality.

One of the first ornithological studies of Asia was undertaken by **Georg Eberhard Rumpf** or **Rumphius** (1627-1702). Having developed a taste for travelling very early, he enlisted in the Venetian army, due to leave for Brazil, but his vessel was captured by the Portuguese. After three years of detention in Portugal, he set sail for Batavia (modern day Jakarta) in June 1653, at the service of the Dutch East

India Company. Despite the instability of the region, he conducted many expeditions on the island. He kept his role as administrator only to feed his family, he would later write. Having become blind, he continued his research with the aid of his son, Paulus Augustus.

Rumphius published his observations in several thematic works on the region of Batavia: *Herbarium Amboinense* is a catalogue of plants, *D'Amboinsche Rariteitkamer* describes the contents of his cabinet of curiosities, *Ambonsche Land-Beschrijving* is a geographic description and *Ambonsche Historie* is a historic work (Ambon is an island in the Moluccas). Death prevented him from completing the part devoted to the fauna that was to be titled *Ambonsch Dierboek*. The latter would be published posthumously through the efforts of his

C . *Het Koninga Vogelen*   D . *De . Amboinsche Land Bek*

Plate taken from *Oud en Nieuw Oost-Indiën* (1724–1726) by François Valentijn, the first realistic representation of a bird of paradise (above). He took his descriptions from the observations of his friend Georg Eberhard Rumph (1628-1702), without crediting him.

friend **François Valentijn** (1666–1727), who appropriated and published it under his own name in the third volume of his vast encyclopaedia *Oud en Nieuw Oost-Indiën* in 1726. Certain sentences reveal their true author, like a comment on the taste of the flesh of the cassowary that could only have been written by Rumphius. We find the description of 41 species of birds, of which the first is a bird of paradise.

A final discoverer of birds; about whom we know very little is **Friderich Martens**. This surgeon-barber embarked on a whaling vessel for an expedition that would bring him to Spitsbergen in 1671. The account of his voyage was published four years later, titled *Friderich Martens vom Hamburg Spitzbergische oder Grönländische Reise Beschreibung gethan im Jahre 1671*. Martens described 15 species of birds observed in the great north. His work was translated into a number of

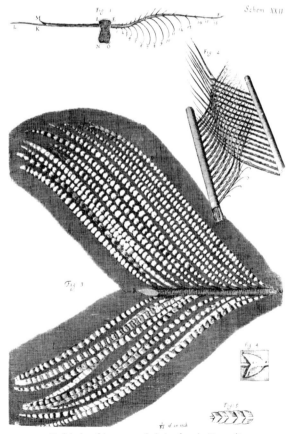

Plate showing feather barbules taken from *Micrographia* (1665) by Robert Hooke.

Hooke's plate of the structure of a feather, from the *Micrographia*.

European languages and would serve as a reference on the fauna of Northern Europe for almost two centuries.

Advances in biology and the permanence of the past

From the point of view of ornithology, this century was a period of transition. The knowledge inherited from antiquity and improved upon during the Renaissance was gradually stripped of its most blatant errors, classifications were refined, collections were organized along more specialist lines, and the techniques of observation were revolutionized.

The microscope allowed the discovery of a world that up until then remained unknown. **Robert Hooke** (1635–1703) studied the attachment of distal and proximal barbules to one another in a goose feathers. He also studied the structure of the iridescent barbules in the feathers of the Peacock.

Amongst the most notable advances in biology, the anatomical work of **William Harvey** (1578–1657) must also be mentioned. He published *Exercitationes de generatione animalium* in 1651, in which he studied the embryonic development of chicken eggs and organs. This would also be the subject of two publications by **Marcello Malpighi** (1628–

Frontispiece (below) of the catalogue of Ole Worm's (above) cabinet of curiosities, *Museum Wormianum* (1655). Voyages enriched collections, whose owners above all sought spectacular objects. However, a movement to renounce this dramatic approach would gain momentum, leading to the creation of modern museums.

Above left and below, two illustrations taken from the catalogue of the cabinet of Ole Worm (1655). Above right, an illustration taken from *Ornithologiae* (1599) by Ulisse Aldrovandi.

1694), *De formatione pulli in ova* and *De ovo incubato*, released in 1672.

However, these significant advancements curiously cohabited with vestiges from the past. **Ole Worm** (1588–1654), personal physician of king Christian IV of Denmark, was a good example of this. Like other biologists of his time, he studied embryonic development and opposed certain commonly held beliefs of his period. He demonstrated that the famous unicorn horn brought back by the Nordic voyagers was nothing more than the tusk of a narwhal, and yet justified its use as an antidote.

Worm is best known for his cabinet of curiosities, which he described in 1655 in *Museum Wormianum* (see the previous page). It is interesting to find a vignette showing a rather rudimentary image of an owl (above left). However, a better depiction (above right) was published half a century earlier in 1599 in Ulisse Aldrovandi's *Ornithologiae* (see page 31). However, not all of the illustrations that appeared in Worm's catalogue are so basic, as the image of the Black-throated Diver (*Gavia arctica*) demonstrates (see left). Precise illustration would become a constant requirement during the 18th century, which would see the birth of true ornithology.

The 18th century
The birth of ornithology

The 18th century is often portrayed as an era in which classification was passionately promoted, which was particularly true in ornithology. Three characters dominated the discipline in this period. The Swedish Linnaeus provided a theoretical framework for the classification of species. His system would become immensely successful, but little remains of his classification above the species level. The Frenchman Brisson followed his system partially; his classification is more refined, but only his descriptions of genera remain. Buffon was opposed to these systems, which he believed ran contrary to nature. For him, species were not fixed. Instead, they consisted of an infinite number of gradations subject to the climate and environment that the Linnaean system could not adequately account for. The unification of Linnaeus's system with Buffon's theories would be achieved by a German, Pallas.

It was also a time when the collective imagination was fascinated with and inspired by nature; Romanticism was not far off. Buffon's texts would immediately prove to be very successful, as would the accounts of great voyages. The popularity of science was stimulated by the sheer number of works that dealt with serious subjects in an accessible way. The time where the same person could be both a mathematician and a philosopher was about to end: the accumulation of knowledge led to greater specialization and marked the beginning of professional ornithology.

Summary

The scientific contribution of private collections in the 18th century

We have seen the role played by cabinets of curiosities in the preceding chapters. Throughout the 18th century, their composition would change and their importance to scientific research would grow (as with the collections of Réaumur and Cabinet of the king). Several key figures would contribute to this evolution.

Albertus Seba (1665–1736) was a renowned collector whose scientific impact was unusually minor. He lived in Amsterdam and like de l'Écluse (see page 38) relished the privilege of being the first to discover plants and animals, living or dead, brought back on vessels returning from distant lands.

Portrait of Albertus Seba (1665–1736), owner of one of the most prestigious cabinets of curiosities of his time.

Plate taken from the catalogue of Albertus Seba's collections (1734–1765).

His collection was so famous that it attracted the attention of Peter the Great (1672–1725) who managed to acquire it around 1716, along with that of his compatriot, Frederik Ruysch (1638–1731). Seba immediately took on the building of a new collection that would soon surpass the previous one. He published a catalogue of the plants, animals and monsters in his possession in *Locupletissimi rerum naturalium thesauri accurata descriptio* (1734–1765). This work was beautifully illustrated and contained only brief explanatory texts. Some of the species of birds that were featured in it are not identifiable with a high degree of certainty, to the point that they are sometimes thought to represent extinct species. The catalogue would be severely criticized. One such critic was Coenraad Jacob Temminck (1778–1858) who, in the introduction of his *Manuel d'ornithologie* (1820), stated that Seba is a "collector without taste and without observational talent, and moreover an imprecise designer".

The name of **Sir Hans Sloane** (1660–1753) is intimately tied to the history of the British Museum in London. A student of John Ray, he left England to spend several years in Jamaica (1687–1689) and studied its flora and fauna. He brought back 800 new species of plants. He also participated in the introduction of chocolate to England, but it was, more

Birds of paradise depicted in the catalogue of Seba's cabinet: *Locupletissimi rerum naturalium thesauri* (1734–1765). It was long questioned whether these bird had legs since the specimens received from New Guinea did not have any. They were prepared by the natives following traditional methods which removed the legs and sometimes the wings. The secrets of these preparations were discovered by Louis Jean Pierre Vieillot (1748–1831) around 1820.

Sir Hans Sloane (1660–1753) gathered an extraordinary collection of 50,000 books and manuscripts, 23,000 coins, and more than 8,000 quadrupeds.

prosaically, thanks to his medical activities in London's high society that he made his fortune. He had started to assemble a large natural history collection in the Caribbean, and upon his return to Great Britain he tirelessly kept improving it. He resorted to dealing with collectors who set out for the newly acquired British possessions, as was the case with Mark Catesby (see page 63). Concerned for the survival of his collections, he bequeathed them to the British crown for the sum of £20,000, a fifth of their actual value, with the obligation to create a grand museum (which would only open its doors in 1759), the future British Museum.

The ornithological portion of his collection comprised 1,172 objects (skins, skeletons, nests and eggs). This number, which may appear rather small to modern readers, can be explained by the great challenges encountered in the preservation of the remains of birds. It must be noted that none of these ornithological specimens have survived. Sloane's importance comes from the fact that he ruled the powerful Royal Society that he directed after the death of Newton in 1727. This great patron of British science played an important part in the development of ornithology, through the expeditions that he organized, the museum that he founded and maintained and through the support that he provided to numerous scientists.

Not all collections had the good fortune of becoming the core of an institution as prestigious as a great museum. In fact, some important collections would succumb to much sadder fates. **Sir Ashton Lever** (1729–1788) deserves a place amongst the great collectors of the 18th century. He began to collect shells around 1760, then fossils, thousands of stuffed birds, and ethnological objects, without giving the slightest clues as to the reasons for his interest. Soon, his collections filled several rooms and required 1,300 showcases. He opened his collections to the public in order to gather the necessary funds to further augment his possessions. These efforts were very successful and attracted a number of scientists, such as Johann Reinhold Forster (1729–1798) who came to study them. However, Lever did not succeed in meeting the escalating costs of maintaining his collections. Near bankruptcy in 1781, he attempted to sell his collections to the British government, who offered him a ludicrous sum that amounted to only a minuscule fraction of their value. Lever was thus required to sell his collections by means of a lottery, which would lead to yet another disappointment when only a quarter of the tickets were sold. These compounding failures had ill effects on his health and he died soon after in great destitution. The happy winner of this lottery, a certain James Parkinson, did not have

The evolution of taxidermy

Taxidermy was essential to the development of bird studies. It became widespread over the centuries, because it tackled a major problem: how to protect the skins of birds from carrion-eating insects. Over the centuries these insects wreaked havoc, ravaging the collections of Sloane, Buffon, and those assembled by James Cook, which have nearly or completely disappeared. Réaumur took an interest in taxidermy and published a text in 1748 in which he proposed several methods of preservation using sulphur, but they proved not very effective.

It was Jean-Baptiste Bécœur (1718-1777), a pharmacist from Metz, who came up with a method based on a blend of arsenic,

soap, potassium carbonate and quicklime. He initially kept his recipe secret in the hopes of commercializing it and only gave it to some of his students like Levaillant. Louis Dufresne (1752-1832), taxidermist at the Muséum de Paris, who learned the recipe from Levaillant, popularized the recipe in an article on taxidermy that appeared in *Nouveau dictionnaire d'histoire naturelle* (1803-1804). It would be used until the present day, but the use of arsenic is not without its dangers and some ornithologists such as Osbert Salvin (1835-1898) lost their lives to its use, while others, like John Cassin (1813-1869) and John Kirk Townsend (1809-1851), had their lives shortened.

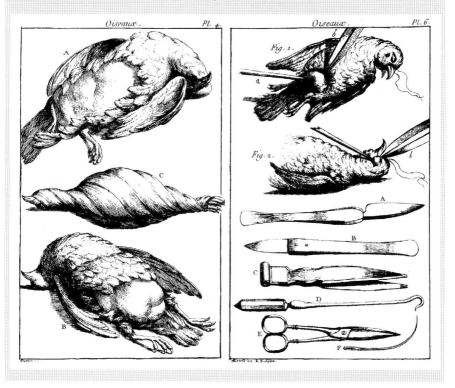

the means to maintain the collection and attempted to sell it in its entireity at a reduced price to the government in 1806. Sir Joseph Banks, president of the Royal Society, who was known to detest Lever, vetoed the move. Finally, the collection was divided into more than 7,000 lots for an auction that would last two months. Though some of the most important birds

This *Réunion d'oiseaux étrangers places dans différentes caisses* painted by Alexandre Isidore Leroy de Barde (1777-1828), shows the evolution of natural history collections in the 18th century. The presentation still seeks to be spectacular, but the birds are depicted in more natural poses.

Marmaduke Tunstall (1743-1790) assembled a vast ornithological collection used by many British ornithologists.

were acquired by the great naturalists of the period or by the Museum of Vienna (such as the Lord Howe Swamphen, *Porphyrio albus*, or the Mascarene Parrot, *Mascarinus mascarin*, two now extinct species), the collection is now nearly completely lost. It was as great a loss for science that certain species were only known from the specimens gathered by Lever. Forever lost among these acquisitions were the birds collected during the voyages of James Cook.

If Seba's approach at the beginning of the century was unscientific, a few decades later the situation would be quite different. The collector **Marmaduke Tunstall** (1743–1790) was a good example. He assembled a large collection, of which the live and stuffed birds alone were valued at £5,000, a veritable fortune at the time. He released a catalogue of his birds titled *Ornithologia britannica* in 1771. Adept at the Linnaean method, he named several species including the

Peregrine Falcon (*Falco peregrinus*), the Grey Wagtail (*Motacilla cinerea*) and the American Pipit (*Anthus rubescens rubescens*). His collections were the object of other studies led notably by Thomas Bewick (see page 87) and Thomas Pennant (see page 82). As we shall see, progress in the preservation of skins would contribute to the incredible popularity of bird collecting throughout the following century.

In a text from 1790, *Mémoire sur les cabinets d'histoire naturelle et particulièrement sur celui du jardin des plantes,* Lamarck (1744–1829) clearly explained the new requirements of scientific collections: "We often see, in effect, natural history collections whose primary objective, of sorts, is to create a spectacle and perhaps to offer an idea of the wealth and luxury of the owner. The state and order of these collections are more suited to decoration and ornament. ... In fact, I would go as far as to say that the collections that I have mentioned are good for nothing." Lamarck listed the points that contribute to the scientific value of collections: firstly he suggested that "the objects that make up these collections must be arranged in proper methodical or systematic order" going from kingdom to species, that "each of them must be identified with certainty", that the specimens must be in good condition and "never disfigured, modified or embellished by art", that a catalogue must establish the list and, finally, that the cabinet be completed by a library.

The evolution of ornithological illustration: Catesby, Albin and Edwards

Illustration has always played an important role in ornithological works, but in the 18th century the nature of illustrations changed and they took on growing importance. Three names illustrated this evolution: Mark Catesby, Eleazar Albin and George Edwards.

Mark Catesby (*c.*1683–1749) studied natural history in London before following his sister and brother-in-law who were emigrating to Virginia in 1712. He stayed seven years and frequently explored the region to study its plants and residents. He participated in an expedition to Bermuda and Jamaica in 1714. Catesby returned to England in 1719 bearing seeds and natural specimens that he then passed on to various British naturalists, among them Samuel Dale (1659–1739) and William Sherard (1659–1728), whom he had met in 1694 through John Ray. In 1722, W. Sherard recommended that Catesby, because of his talents as an illustrator and his experience in natural history, should organize a scientific

Title page of *The Natural History of Carolina, Florida and the Bahama Islands* (1731–1748) by Mark Catesby.

The American Flamingo (*Phoenicopterus ruber*), copper engraving, hand-painted by Mark Catesby, taken from *Natural History of Carolina, Florida and the Bahama Islands* (1731–1748).

mission in the Carolinas for the Royal Society. The voyage was financed by a number of naturalists, among them Sir Hans Sloane (1660–1753).

Catesby spent three years exploring North and South Carolina, Georgia and Florida. He also set out to study fish in the Bahamas and took an interest in the American Indian peoples that he encountered. He keenly observed the impact of the seasons on the fauna and flora, and sent numerous specimens of new species to the Royal Society. Upon his return to Great Britain in 1726 he learned engraving techniques from French artist Joseph Goupy (1689–1769). He devoted more than 10 years to his primary work, *The Natural History of Carolina, Florida and the Bahama Islands*, published in English and in French from 1731 to 1748. Of the 220 plates that he completed himself, 109 depict birds. In the text, he provides explanations of the habits and behaviour of the species, but also borrowed passages from *New Voyage to Carolina* (1709) by John Lawson (*c.*1674–1711). It is thus that Catesby completed the first scientific study of the fauna and flora of North America. He became a member of the Royal Society in 1733 and published an essay the following year in which he refuted the assertion that birds wintered under marsh waters. His descriptions would be taken up by numerous authors, including Carl Linnaeus.

Much of the detail of the life of **Eleazar Albin** is in doubt, including his date of birth (probably 1680). A professional watercolour artist by trade, he also taught drawing and painting. He is best known for his work as an illustrator of naturalist works. Albin first took an interest in insects and the oldest preserved examples of his drawings depicting these are from 1711. He met a number of naturalists in this period who would entrust him with the illustration of their entomological works, notably Sir Hans Sloane (1660–1753). Encouraged by his clients, he published a book on insects in order to pique the public's interest and to promote its study. Despite the

Eleazar Albin, by Jean-Baptiste Scotin (detail from the cover of *Natural History of Spiders*, published in 1736).

18TH CENTURY

The Hoopoe (*Upupa epops*), a coloured plate from *A Natural History of Birds* by Eleazar Albin.

difficulties that he encountered in acquiring subscriptions, *A Natural History of English Insects* (illustrated with about 100 plates) was released in 1720.

He followed this work with three volumes devoted to birds in 1731, 1734 and 1738, titled *A Natural History of Birds*. This work contained 306 hand-coloured copper engravings. The quality of the illustrations was highly variable: certain birds are not recognizable, and it has been suggested that they may represent extinct species. Though he claimed to have drawn from live specimens, some are clearly drawn from dead birds. His daughter, Elizabeth, and his son, Fortin, also assisted with the illustrations.

Numerous birds are described here for the first time, including the Tree Pipit (*Anthus trivialis*), the Common Greenshank (*Tringa nebularia*), the Bearded Tit (*Panurus biarmicus*) and the Golden Pheasant (*Chrysolophus pictus*). Nonetheless, Albin's scientific understanding is manifestly mediocre and he draws his core text from Willughby's *Ornithologiae*, not adding any truly new insights.

Albin's work achieved a good measure of success, which led its author to publish further works, including a small volume entitled *A Natural History of English Song Birds* in 1737. The 23 plates depicted male and female as well as the egg for each species. It was the first time a naturalist artist had used this type of presentation. The work was reprinted twice (in 1741 and in 1759) and was also released in pirated editions in 1754, 1791 and 1825. Albin died in 1741 or 1742.

George Edwards (1693–1773) was born of a well-to-do family and his father hoped he would take up a career in business. He devoted his leisure time to the study of a wide range of subjects including astronomy, natural history and painting. His schooling ended in 1718, the year he left for the Netherlands, then for other parts of continental Europe. Upon his return to England two years later, he earned some money creating illustrations of natural history for various members of the Royal Society, and proved to be particularly skilled in illustrating birds. He returned to the continent in 1731 where he became interested in ancient works of natural history and the paintings of the great masters.

On his return to England, he was chosen by Sir Hans Sloane as the librarian of the Royal College of Physicians in London. Sloane put him in charge of painting the specimens of his private museum, among other things. Edwards appreciated the accomplishment of the illustrations in Albin's works, but thought they lacked vitality. With this in mind, he threw himself into the publication of *A Natural History of Uncommon Birds*, the first volume of which appeared in 1743 and the last

It was at 40 years of age that George Edwards (1693-1773) published his first illustrated natural history works.

The Greater Frigatebird (*Fregata minor*) by George Edwards in *Gleanings of Natural History* (1758-1764). He sought to produce more vivid illustrations than those of his predecessors.

Title page of *Gleanings of Natural History* (1758-1764) by George Edwards.

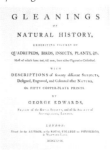

in 1751. Carl Linnaeus, who met Edwards during his trip to London in 1736, would make use of the English names of new species given by Edwards by translating them into Latin.

The illustrations were drawn from live animals, as Edwards asserted in his introduction. In spite of his efforts, many of the illustrations lack realism and the birds appear somewhat inanimate. He often places other animals (like an insect, a mammal, a fish or a lizard) next to the birds. The success of his book drove Edwards to pursue other work and he published *Gleanings of Natural History* from 1758 to 1764. Though the title does not suggest continuity, it is in fact an extension of his first publication. He again describes numerous British species, but exotic birds are more numerous. To obtain specimens, he kept a large network of contacts, including William Bartram (see page 95). He described a number of new species like the White-throated Sparrow (*Zonotrichia albicollis*) and the Yellow-throated Warbler (*Dendroica dominica*), two species from the Americas. His descriptions would be, by and large, taken up by Linnaeus in his *Systema Naturae*.

Frontispiece of *A Natural History of Uncommon Birds* (1743-1751) by George Edwards. This work brought such renown to its author that he would be named the father of British ornithology.

Following his studies, the Austrian **Johann Ferdinand Adam von Pernau, Lord of Rosenau** (1660-1731), travelled extensively throughout Europe and visited Italy, France and the Netherlands. In 1690, he settled near Coburg in Bavaria. He developed a passionate interest in birds and, unlike many ornithologists of his time, he did not content himself with studying them in the calm of a cabinet, and observed many in nature. While breeding wild species, Pernau noticed that certain individuals would become attached to him, and once set free, would return to him of their own volition. This insight on bird behaviour cannot be mentioned without bringing up the work of Konrad Lorenz (see page 201) three centuries later on the phenomenon of imprinting. Pernau anonymously published his observations in 1702, and they would be reprinted in 1707 and 1716.

In another text from 1720, Pernau remarked that bird songs are not necessarily instinctive, suggesting that they are in fact sometimes learned. He also asserted that migrations are not caused by changes in temperature or by the search for food, but rather that they are governed by an as yet unknown factor. He questioned the notion of territoriality in birds. His research was very modern and made him one of the fathers of ethology.

He would pen the first known written statement of contempt for the killing of birds: "It is not my intention to describe how to capture birds ... but to describe the pleasure of observing these beautiful creatures of God without killing them."

Pernau would greatly influence **Johann Heinrich Zorn** (1698-1748), a German pastor, who in his book *Petino-Theologie, oder Versuch die Menschen durch naehere Betrachtung der Voegel zur Bewunderung, Liebe... ihres mächtigsten... Schöpffers aufzumuntern...* (1742-1743) sought to provoke admiration and respect for the divine power through the observation of birds. He carried out remarkable and precise behavioural observation, like Pernau before him. Zorn observed that the colour of birds or eggs often corresponds to that of their environment so as to be better camouflaged. No one would take up Pernau and Zorn's research until the 20th century, when the first true scientific studies on bird behaviour would appear.

Réaumur's private collection and the work of Brisson

René-Antoine Ferchault de Réaumur (1683–1757) did not contribute directly to ornithology. His name is primarily associated with physics experiments and entomology. His *Mémoires pour servir à l'Histoire des insectes* (1734–1742) marked an essential step forward for this area of science. His only personal research on birds dealt with methods of artificial incubation of chicken eggs, as well as digestion in granivorous and carnivorous birds.

It was his private cabinet of curiosities, one of the most well endowed in Europe, that would allow for a major contribution to ornithology in this century. Réaumur, thanks to his duties at the Académie des Sciences, had an important network of contacts at his disposal, including Pierre Poivre (1719–1786) on the island of Réunion, Michel Adanson (1727–1806) in Senegal, Charles de Geer (1720-1778) in Sweden, and Jacques-François Artur (1708–1779) in French Guiana. Réaumur encouraged them to send him specimens, but also, as often as possible, to provide additional details on the biology of the species and to procure nests and eggs. To facilitate the sending of parcels from his correspondents, he obtained, thanks to his relationship with the postal superintendent, paid delivery for the parcels sent. He also presented a recipe for the preservation of birds to the Académie des Sciences that would prove to be rather ineffective.

From 1749, Réaumur entrusted the conservation of his collections to one of his young relatives by marriage, **Mathurin Jacques Brisson** (1723–1806). Before having assumed the responsibility of Réaumur's cabinet, Brisson wanted to join the church. He had been appointed a deacon in 1747, but his passion for science led him to give up this position.

Brisson released one of the major ornithological works of the 18th century, whose title states its purpose: *Ornithologie ou méthode contenant la division des oiseaux en ordres, sections, genres, espèces et leurs variétés, à laquelle on a joint une description exacte de chaque espèce, avec les citations des auteurs qui en ont traité, les noms qu'ils leur ont donnés, ceux que leur ont donnés les différentes nations, & les noms vulgaires* (1760). It consisted of six volumes and some 4,000 pages. This first catalogue of all the world's species of birds would be a model for a great many authors for more than a century.

Thanks to his duties as a curator, Mathurin Jacques Brisson had access to one of the largest collections of birds ever assembled. He could also gain access to other private collections

René-Antoine Ferchault de Réaumur (1683–1757), an important figure in the scientific life of the 18th century.

Mathurin Jacques Brisson (1723–1806), conservator of Réaumur's collection.

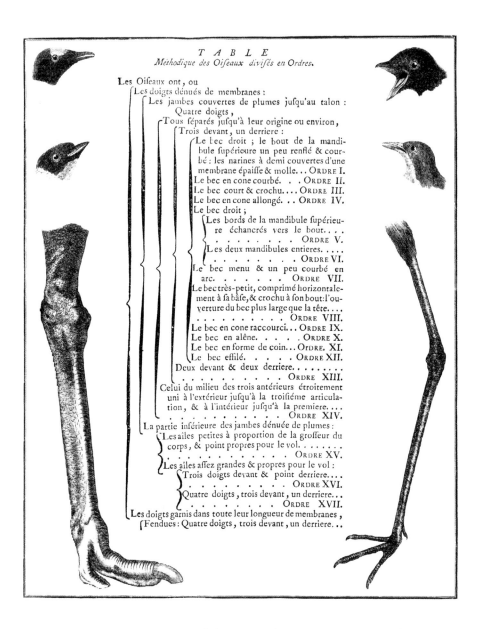

Page taken from *Ornithologie* (1760) by Mathurin Jacques Brisson.

in Paris and the surrounding area such as those of Pierre Jean Claude Mauduyt de la Varenne (*c.*1732–1792) or of the Aubry abbey. Mathurin Jacques Brisson's work is above all the catalogue of a particular collection. That is why a description of a teratological specimen of a chicken with five digits can be found among the others. For the same reason, Brisson did not take an interest in the anatomy of the species he described – because he worked from dried animals, he could only make

A hornbill, illustration taken from *Ornithologie* by Mathurin Jacques Brisson, engraved by François-Nicolas Martinet (*c.*1760–*c.*1800), who would also sign Buffon's engravings.

observations based on the skins – and as such, details on behaviour, biology or ecology are hardly ever given.

Brisson completely rethought the classification proposed by Jacob Theodor Klein (see page 84) and particularly by Linnaeus (see page 78). While the Swedish scholar had proposed vast disparate genera grouped in only six orders and 51 genera, Brisson created 115 genera grouped into 26 orders. He described 320 new species of a total of approximately 1,500. Contrary to Linnaeus, who favoured as brief descriptions as possible, Brisson took care to provide the greatest amount of detail. When Brisson's work began to appear, he did not have a copy of the 10th edition of Carl Linnaeus's *Systema naturae*, but only that of the sixth edition in which the binomial system had not yet been cemented. This explains why, despite the great merit of Brisson's classification, only his names of genera have been retained and not his species names. Brisson was devoted to establishing simple and obvious criteria and avoiding too much complexity, which he believed would lead to more errors.

His work had a considerable impact and his classification, in spite of being more difficult to learn than that of Linnaeus, would be taken up by John Latham (1740–1837) in *A General Synopsis of Birds* (1781–1801), by Pierre Jean Claude Mauduyt de la Varenne (*c.*1732–1792) in the ornithological article of *Encyclopédie méthodique* (1783), by Thomas Pennant (1726–1798) in *The British Zoology* (1766) and by Pierre Sonnerat (1748–1814), nephew of Pierre Poivre (1719–1786), in *Voyage aux Indes orientales et à la Chine* (1782). Brisson's *Ornithologie* was reprinted twice during his lifetime, but it would not achieve the same level of success as Buffon's *Histoire naturelle des oiseaux*.

Frontispiece of *Ornithologie* by Mathurin Jacques Brisson. It was its author's only work on birds.

Unfortunately for Brisson, and perhaps for the history of ornithology, his research on birds was abruptly interrupted. Réaumur died in 1757 and although he had bequeathed his collections to the Académie des Sciences, they were intercepted by Buffon who had them transferred to the cabinet of curiosities at the Jardin du Roi that he oversaw. Brisson therefore lost his position and all hopes of being able to continue to work in this area. Discouraged, on the advice of abbot Jean-Antoine Nollet (1700–1770) he turned to physics, which he taught at the Collège de Navarre in Paris, and completely ceased his ornithological research. Buffon's behaviour may be justified by the fact that, unlike the Jardin du Roi, the Académie des Sciences did not have the means to look after this sort of collection. However, the fact remains that 18th century ornithology prematurely lost one of its most important figures. Brisson's last influence on zoology would be to convince one of his students, Étienne Geoffroy Saint-Hilaire (1772–1844), to devote himself to the study of animals.

Buffon's Histoire naturelle, the man of the Jardin du Roi

Son of a Burgundian councillor, **Georges-Louis Leclerc, Count of Buffon** (1707–1788) studied at Dijon. Having graduated in law in 1726, he abandoned judicature, the career his father no doubt wished for him, to devote himself to science and philosophy. Buffon was particularly drawn to mathematics at this time. For a few years he led a carefree life and travelled in France and Italy before settling in Paris in 1732. A clash would put him at odds with his father. The latter, widowed a year earlier, wanted to remarry. Buffon, fearing for the fortune left by his mother, threatened his father with a lawsuit. In turn, he received his maternal inheritance and bought the Buffon family land (located fairly close to Montbard, his birthplace). In 1733, he presented his first paper to the Académie des Sciences in which he solved a mathematical problem that would win him the praise of academics. In 1734, the year of his entry into the Académie des Sciences as an adjunct-engineer, Georges-Louis Leclerc took on the name Buffon. He founded a nursery at Montbard and began to conduct experiments on the resistance of wood, which earned him the attention of the king and a pension of £2,000. In 1739, Buffon left the engineering division of the Académie for the botany division. He succeeded, thanks to his good relationship with the king, to be appointed head of the Jardin du Roi (which is now the Muséum National d'Histoire Naturelle) in the place of the established botanist

Georges-Louis Leclerc de Buffon (1707-1788) is one of the most prominent figures in 18th century natural history. He was the author of a bestselling work, and was behind the future Muséum d'histoire naturelle in Paris's rise to prominence. He showed a great deal of hostility towards Réaumur and Linnaeus.

Duhamel du Monceau (1700–1782), who accused Buffon of plagiarising a treatise on wood and techniques applied to it. This nomination unleashed much controversy, but he would retain the position for 50 years.

Buffon began to reorganize the Jardin du Roi while continuing to further his career. In 1744, he was appointed perpetual treasurer of the Académie des Sciences, a position that included a pension of £3,000. He entered into the Académie Française in 1753 and directed it in 1760.

With the Jardin du Roi, Buffon inherited one of the oldest scientific institutions, founded in 1635. Thanks to his work

The Wild Turkey (*Meleagris gallopavo*), plate taken from *Histoire naturelle* by Buffon. Scientific names did not appear in the original edition, but they were later added by hand.

LE DINDON.
MELEAGRIS GALLO-PAVO.

The hummingbird in Buffon's *Histoire naturelle*

Of all animate beings, it is the most elegant in form, and the most brilliant in colour. The stone and metals polished by the art of man are not comparable to these jewels of nature; nature has placed it in the order of birds, lowest on the scale of size, *maxime miranda in minimis*; its masterpiece is the small hummingbird; it has been bestowed with all the gifts of other birds, light-ness, rapidity, agility, grace and rich finery, all belong to this little favourite. The emerald, ruby and topaz shine on its regalia; it never soils them with dust from the earth, and in its entirely aerial life we barely see it touch down to the grass for the briefest instant; it is always in the air, flying from flower to flower; it shares their freshness and their radiance: it lives of their nectar and only resides in climates where they never cease to renew them-selves. It is in the warmest parts of the New World that all of the species of hummingbirds can be found; they are quite numerous and remain confined between the two tropics. Those that set out for temperate regions in the summer make it a short stay; they seem to follow the sun, advancing and with-drawing with it, carried on the wings of zephyrs in pursuit of the eternal spring.

The Indians, struck by the radiance and luminosity of the brilliant colours of these birds, named them rays or hair of the sun. The Spanish called them *tomineos*, a word referring to their excessive smallness. The tomine is a weight of 12 grains: I have seen, said Nieremberg, one these birds weighed in a bird-trap, which with its nest weighed no more than two *tomines*, and as for the volume of the smallest species of these birds, they are smaller than the large horsefly and the bumblebee in size ...

Nothing equals the vivacity of these small birds, if not their courage, or rather their audacity: they can be seen in furious pursuit of birds 20 times larger than them, attaching themselves to their bodies, and allowing themselves to be carried by their flight, pecking at them repeatedly,

Ruby-throated Hummingbird, from Cayenne

until they have calmed their little rage. From time to time, they even engage in combat with each other; impatience seems to characterize their spirit: if they approach a flower and find it to be wilted, they remove its petals with haste characterizing their scorn; they have no other voice other than a small cry, *screep, screep*, frequent and repeated; they make themselves heard in the woods from dawn, until the first rays of sunlight, when they all take flight and disperse themselves throughout the countryside.

They are solitary creatures, and it would seem difficult, always being up in the air, to recognize each other and pair up; nevertheless, love, whose power extents beyond that of the elements, brings together and unites these scattered beings; we see hummingbirds in pairs during the nesting phase: the nest is constructed to accommodate their delicate bodies; it is made of a fine cotton and a silky down gathered from flowers; this nest is tightly woven, the consistency of soft, thick skin; the female takes care of the assembly, and leaves the gathering of the materials to the male; we see her attentive care in this cherished work, searching for, choosing and incorporating the fibres bit by bit to form the tissue of the soft cradle that will bear her progeny; she polishes the sides with her throat, the inside with her tail; she lines the exterior with small pieces of gum tree bark that she glues around it to defend it from the abuses of the air, as much as to make it more solid; the whole nest is attached to two leaves or a single twig of an orange or lemon tree, or at times a wisp of straw hanging from the roof of a building. This nest is no larger than half an apricot, and of the same shape; we find two white eggs, no larger than small peas; the male and the female take turns sitting on the nest for 12 days; the young hatch on the 13th day and are no larger than flies.

this garden, which had been in decline, gained considerable renown. He increased its acreage by acquiring neighbouring lots, sometimes violently expelling landowners. He improved the collections of the small cabinet of curiosities in the Jardin's possession to create a luxurious display, lauded in the guides printed for the public. To do so, he obtained the collections of other scholars of his time, such as the anatomical pieces of Joseph Bonnier de La Mosson (1702–1744) and, in 1758, Réaumur's cabinet.

Like Réaumur, Buffon had a vast network of contacts at his disposal (often the same contacts as Réaumur). They would send seeds and plants for the Jardin du Roi, but also live and stuffed animals to be studied by Buffon and other naturalists employed by the Jardin. In order to reward donations without dispersing funds, he created a certificate for the correspondents of the Jardin du Roi. His notoriety became such that even monarchs contributed to enriching its collections, including the kings of Denmark, Poland and Prussia, and the Empress of Russia.

Buffon also had a knack for hiring valuable members of staff such as gardener-botanist André Thouin (1747–1824), botanist Antoine-Laurent de Jussieu (1748–1836), chemists Pierre Joseph Macquer (1718–1784) and Antoine François Fourcroy (1755–1809), anatomists Antoine Ferrein (1693–1769) and Louis Daubenton (1716–1800), botanist and zoologist Jean-Baptiste de Lamarck (1744–1829), zoologist Lacépède (1756–1825) and geologist Barthélemy Faujas de Saint-Fond (1741–1819).

During the same period, Buffon began work on a vast project, *Histoire naturelle générale et particulière: avec la description du Cabinet du Roy*, the first volume of which would be released in 1749. His ambition was to describe the whole of the natural world, from minerals to human beings, which he considered to be like an animal "with its habits varying according to races and climates". The success of this work was immediate: the print run of the first volumes, which contained descriptions of the cabinet of the king, was exhausted in a few weeks. Translation into German began in 1750.

In 1766, having completed a history of quadrupeds, he commenced the writing of *Histoire naturelle des oiseaux*, which would appear in nine volumes from 1770 to 1783. A deluxe illuminated edition was also published from 1771 to 1786 in ten volumes. His *Histoire naturelle* was a veritable bestseller. It would be released in a great number of subsequent editions in full or in compilations, illustrated or not. The texts of *Histoire naturelle des oiseaux* were written primarily by Philippe Guéneau de Montbeillard (1720–1785), and from the sixth volume by

Title page of *Histoire naturelle des oiseaux* which was a part of Buffon's *Histoire naturelle générale et particulière* published in 36 volumes.

HISTOIRE
NATURELLE
DES OISEAUX.

Tome Second.

A PARIS,
DE L'IMPRIMERIE ROYALE.
M. DCCLXXI.

abbot Gabriel Bexon (1748–1784). Buffon supervised their publication and readily edited them. Buffon omitted anatomical descriptions by Louis Daubenton, which he judged would be off-putting to his readers. This would unleash a long quarrel between the two men.

Like Brisson's work, Buffon's was closely connected to a great collection. Buffon did not cease his efforts to enrich the collections at his disposal at the Jardin du Roi. He believed that a species could not be adequately described unless an example of a male, female and offspring were available to be studied. When *Histoire naturelle* began to appear in 1749, he had only 60 to 80 specimens of birds, many of which were in very poor condition. The collections included about 800 species 20 years later, notably thanks to the addition of Réaumur's collection. Around 1775, he received 160 species, most of them new, gathered by Charles Nicolas Sigisbert Sonnini de Manoncourt (1751–1812) during his three-year voyage to French Guiana.

Though only 2,000 species of birds were known at the time, Buffon would never meet his objective of acquiring several specimens of each species. He would have had to gather about 8,000 specimens, ten times what he had collected by 1770.

When it came to classification, Buffon expressed great disdain for Linnaeus's system, which he thought too artificial. He believed that an understanding of the habits and behaviour of species was essential in describing them and that the notion of a fixed species promoted by Linnaeus was far removed from reality. He attacked the Linnaean system from the first volume of his *Histoire naturelle*. Buffon asserted that it was impossible to apply a general system to classify different forms of life because "nature operates in unknown gradations, and consequently she does not lend herself to these divisions [species, genera, families, etc.], since she passes from one species to another, and often one genus to another, through imperceptible nuances". He did not, however, necessarily deny the existence of the species. He provided a definition very close to our own, based on interbreeding (two individuals are of the same species if their descendants are fertile). According to him, the differences between individuals were the result, for example, of the influence of climate. Thus, Buffon assembled related species under a single name: he regarded the Great Grey Shrike (*Lanius excubitor*) and the Lesser Grey Shrike (*Lanius minor*), considered to be distinct species by the ornithologists at the time, as one species whose appearance varied based on habitat. He made other such connections, a move that would attract some critics. Pierre Sonnerat (1748–1814), in his *Voyage à la Nouvelle-Guinée* (1776), highlighted the fact that parrots constituted "perhaps the most varied ... genus of birds" because

MARTIN PÊCHEUR.
ALCEDO ISPIDA.

MARTIN PÊCHEUR DU SÉNÉGAL.
martinet.   ALCEDO RUDIS? Gm.

each of the islands of the Filipino archipelago "is home to one or several species of this genus, that are their own, and that one does not find on the other islands of the same archipelago, however small the distance between them may be." Sonnerat contradicted Buffon's hypothesis by remarking that the climate and diet were identical in these different locations, and he concluded: "Another explanation must be found, and I leave the discussion to naturalists..." Buffon's opposition to the systems would gradually weaken with time as different volumes of *Histoire naturelle* were published, and he eventually described birds by genera rather than species.

The Common Kingfisher (*Alcedo atthis*) and Pied Kingfisher (*Ceryle rudis*), illustrations taken from Buffon's *Histoire naturelle des oiseaux*.

The first part of *Histoire naturelle* on birds was published in 1770, ten years after Brisson's *Ornithologie*. It included 1,008 plates (973 devoted to birds, representing 1,239 species) coloured by hand, published as *Planches enluminées* (illuminated plates). They were the work of François-Nicolas Martinet, working under the direction of Edme-Louis Daubenton (1732–1786). With the publication of *Tables des planches enluminées* by Pieter Boddaert (1730–*c*.1796) in 1783, connections could be made between Buffon's legends and Linnaean scientific names. These tables also provided an identification key for species.

It is undeniable that a great deal of Buffon's success can be attributed to his style. He proscribed daunting anatomical descriptions and presented the habits and the nature of animals in a lively and often imaginative way. His style would draw harsh criticizm from naturalists who saw him as nothing but a hack. It is no doubt due to its stylistic qualities that Buffon's *Histoire naturelle des oiseaux* was much more popular than Brisson's *Ornithologie*, despite the greater scientific impact of the latter. However, the fact remains that Buffon would have an immense influence on generations of readers.

Carl Linnaeus: a system that changed the face of natural history

The most striking progress in field of classification came first with the work of botanists: Pierre Magnol (1638–1715) and particularly Joseph Pitton de Tournefort (1656–1708) who established its guiding principles. In zoology, the first attempt at classification was proposed by John Ray (see page 44). However, it was the work of a Swede that would give rise to the emergence of modern classification.

Carl Linnaeus (1707–1778), was the son of the vicar of the Lutheran church in Råshult. The son and grandson of pastors, the young Linnaeus displayed a keen interest in botany. He left to attend Uppsala University in 1728, enrolling in medical studies, which included courses in botany, the real motivation behind his choice. To earn a living during his studies, he became the private tutor to the sons of Olof Rudbeck the Younger (1660–1740), physician and naturalist, who allowed him to make use of his library. From 1730, he gave his first botany lectures which proved very popular, attracting up to 400 people. During his stay he improved the university's botanical gardens, and also studied the area's birds. Linnaeus left Uppsala in 1732 to travel to Lapland, which allowed him to gather a large number of specimens and observations.

On his return in 1734, he continued his studies and

Portrait of Carl Linnaeus (1707–1778). He thought his *Species Plantarum* to be the "great achievement of the scientific community" and his *Systema Naturae* "a masterpiece that cannot be read or admired enough".

travelled throughout Sweden. The following spring, after becoming engaged, he decided to pursue his medical studies in the Netherlands, which offered more advanced university education than was available in Sweden. In Amsterdam, Linnaeus visited the cabinet of curiosities of Albertus Seba (see page 58), who requested his expertise in creating its catalogue, but Linnaeus refused. He left for Harderwijk, a city near Amsterdam, whose university was renowned for the quick awarding of degrees. After a week's stay and a quick exam, Linnaeus obtained the degree of doctor of medicine that allowed him to complete his studies at Leiden. On arrival there, he paid homage to Herman Boerhaave (1668-1738), the famous physician. Linnaeus's botanical knowledge greatly impressed Boerhaeve and the two men would soon become friends. Linnaeus prolonged his stay in the Netherlands with the support of wealthy English banker George Clifford (1685-1760), a great collector of plants. In return, he carried out the cataloguing of his collection.

After having travelled in England then France, he settled in Stockholm in 1739, where he became a physician and very soon married. He obtained, three years later, the highly coveted position of professor of medicine at Uppsala University. In 1747, he became the physician of the royal court and was ennobled in 1761. In Uppsala, Linnaeus devoted himself to his studies, to tending the botanic garden and particularly to his lectures, given from the stage or, on occasion, as part of his renowned outdoor demonstrations.

- Classification according to Linnaeus

Linnaeus first made contact with Laurens Theodore Gronow, known as Gronovius (1730-1777), during his stay in the Netherlands. Gronow would later finance the publication of the first edition of his *Systema naturae* in 1735, in which Linnaeus would introduce his system of classification for the first time (though only dealing with botany). This first edition was nothing more than a booklet of only 14 pages, but Linnaeus continued to improve and add to his work, and it would see 16 editions, Linnaeus himself supervising 12. The 12th edition was published in three volumes with a total of 2,300 pages.

Linnaeus proposed a denomination based on a pair of names (the genus followed by the species) or binomial. This was not his invention, as many other naturalists used a similar system. For example Gessner distinguished tits with two names: *Parus cristatus* (Crested Tit), *P. major* (Great Tit), *P. ater* (Coal Tit). However, Linnaeus was the first to develop a system of rules, seeking to standardize the work of naturalists. The

Cover of the 10th edition of Linnaeus's *Systema naturae* (1758).

definitions for genera and species and the generalisation of the use of binomials would only appear in the sixth edition of *Systema Naturae* in 1748. Linnaeus' hierarchical system was based on five taxonomic categories: *classis* (class), *ordo* (order), *genus*, *species* and *varietas* (variety).

The first attempt at applying his system to animals can be dated to the publication of *Museum S:ae R:ae M:tis Adolphi Friderici regis* Linnaeus would only include all animals in the 10th edition of *Systema naturae* in 1758. He described through different works, but mainly this 10th edition, about 1,000 species of birds, of which 714 remain valid today.

If the Linnaean system has persisted until today, the classification he originally proposed has almost been entirely abandoned. He had classified mammals and birds into six orders, distributing the latter based on two anatomical criteria, the shape of the beak and that of the feet:

- *Accipitres* (birds of prey and shrikes);
- *Picae* (parrots, woodpeckers, cuckoos, ravens, birds of paradise, kingfishers, etc.);
- *Anseres* (ducks, swans, geese, petrels, gulls, terns, etc.);
- *Gallinae* (gamebirds, Ostrich, bustards, etc.):
- *Passeres* (pigeons, thrushes, larks, hummingbirds, finches, treecreepers, wagtails).

His system would quickly be the object of both enthusiasm and criticism. His opponents included some of the greatest names in natural history of the period such as Buffon in France and Albrecht von Haller (1708-1777) in Germany. The latter blamed Linnaeus for changing all the names of animals: "The limitless domination that he assumes in the matter of the animal kingdom would be heinous to many people. He takes himself for a second Adam, and has named all animals according to their distinctive traits, disregarding his predecessors. He all but makes man a monkey or the monkey a man." Johann Friedrich Blumenbach (1752-1840) rejected Linnaeus's decision to base his classifications on unique criteria: the great German physiologist and anthropologist believed that the whole anatomy of the animal should be taken into consideration. If Linnaeus's botanical knowledge attracted little criticism, his zoology was heavily questioned. Thomas Pennant (see page 82) said that "like ornithology, [Linnaeus] is too superficial for his opinion to be taken into account...".

If his opponents were virulent, his supporters were no less active. One could rightly speak of the emergence of a sort of cult around Linnaeus, which the latter would indulge with evident pleasure. The numerous Linnaean societies that were created throughout the world following his death testify to the lasting quality of his legacy.

This discussion of Linnaeus would not be complete without mentioning his numerous students. Their work would actively further the development of natural history during the end of the 18th century; for example Charles de Géer (1720–1778) and Johan Christian Fabricius (1745–1808) made important contributions to entomology. Linnaeus also encouraged his students, whom he modestly baptised 'apostles', to travel. One of the most famous was the Swedish Daniel Solander (1733–1782) who participated in James Cook's first expedition around the world, as well as Carl Peter Thunberg (1743–1828) who studied the fauna of Japan and Pehr Forsskål (1732–1763) who explored the Middle East. Philibert Commerson (1727–1773), who travelled with Bougainville, maintained strict correspondence with Linnaeus. Five of the 20 'apostles' would not return from their travels; this had a lasting impact on Linnaeus. He took it upon himself to publish their observations and commemorated each of them in species names.

- The progressive adoption of the Linnaean classification in Europe

Abbot **Pierre Joseph Bonnaterre** (c.1752–1804) is known for being the first French ornithologist to break with the Buffonian tradition to adopt the Linnaean system. Bonnaterre wrote the parts devoted to zoology in *Tableau encyclopédique et méthodique des trois règnes de la nature* published by Charles-Joseph Panckoucke (1736–1798) from 1788 to 1792. He created 12 classes subdivided into 112 very variable genera.

Morten Thrane Brünnich (1737–1827) played a prominent role in the development of Danish zoology. He devoted himself first to entomology and released his observations in *Danske Atlas* (1763–1781), edited by Erik Pontoppidan (1698–1764), bishop of Bergen and a naturalist. In 1762, he was given the responsibility of the conservation of two large Danish private collections. Their study would orient him towards ornithology. In 1764, he released *Ornithologia borealis,* in which he described the new species brought from the north of Europe and the North Atlantic by explorers, among them the Great Northern Diver (*Gavia immer*) and the Manx Shearwater (*Puffinus puffinus*). He was appointed a lecturer at the University of Copenhagen and left to study the fish of the Mediterranean in Marseille. An eclectic mind, he would abandon zoology towards the end of his life to devote himself to mineralogy.

The revolution brought by Linnaeus's precise system prompted certain authors to name species, already previously described, with a Linnaean binomial. Such was the case of

M. TH. BRÜNNICHII
ORNITHOLOGIA
BOREALIS,
Sistens
COLLECTIONEM AVIUM
EX OMNIBUS, IMPERIO DANICO SUBJECTIS,
PROVINCIIS INSULISQVE BOREALIBUS
HAFNIÆ FACTAM,
Cum
DESCRIPTIONIBUS NOVARUM,
NOMINIBUS INCOLARUM,
LOCIS NATALIUM
ET
ICONE.

IMPRIMATUR, J. C. KALL.

HAFNIÆ, MDCCLXIV.

Frontispiece of *Ornithologia borealis* (1764) by Morten Thrane Brünnich.

Morten Thrane Brünnich (1737–1827) first turned to theology but Linnaeus's lecture convinced him to devote himself to natural history.

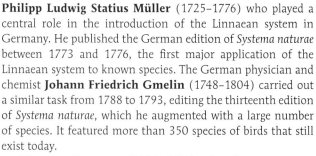

Philipp Ludwig Statius Müller (1725-1776) who played a central role in the introduction of the Linnaean system in Germany. He published the German edition of *Systema naturae* between 1773 and 1776, the first major application of the Linnaean system to known species. The German physician and chemist **Johann Friedrich Gmelin** (1748-1804) carried out a similar task from 1788 to 1793, editing the thirteenth edition of *Systema naturae*, which he augmented with a large number of species. It featured more than 350 species of birds that still exist today.

Thomas Pennant (1726-1798), the first prominent British ornithologist since the death of John Ray, participated in the introduction of Linnaeism in his country. He made his fortune from a rich lead mine he discovered on a property inherited from his father. An active member of regional politics, he was the sheriff of Flintshire.

He discovered natural history at 12 years old during a lecture on Ray and Willughby's *Ornithology*. His penchant for science was encouraged by one of his lecturers, the geologist William Borlase (*c.*1696-1772). He released a great number of books and focused on British fauna as well as that of the Arctic and Asia: *British Zoology* (1761-1766), *Indian Zoology* (1769), *Synopsis of Quadrupeds* (1771) and *Arctic Zoology* (1784-1787) in which he studied the specimens sent to him by Pallas. Pennant was regretful for the slow progress of ornithology in his country. Though he was in correspondence with Linnaeus, he preferred to use only the English names for designating birds. It was under the influence of Johann Reinhold Forster (1729-1798) that he developed, in revising his *Arctic Zoology,* an approach blending of Linnaean classification with that of John Ray.

Like Pennant, British physician **John Latham** (1740-1837) did not initially follow the Linnaean system for new species in his *General Synopsis of Birds* (1781-1801), but he soon realized that posterity would not retain his descriptions if the denominations that he employed did not follow the Linnaean binomial system. Also, he decided to produce an index assigning a Latin binomial to all his species in *Index ornithologicus* (1790-1809). Unfortunately, he was beaten to it by J.F. Gmelin's publication of the 13th edition of Linnaeus's *Systema naturae*. A number of species described by Latham are therefore attributed to Gmelin. In his *Index ornithologicus*, Latham classified birds into terrestrial birds (with six orders: hawks or *accipitres*, magpies or *picae*, passerines or *passeres*, pigeons or *columbae*, gallinaceans or *gallinae* and ostriches or *struthiones*) and aquatic birds (with three orders: wading birds or *grallae*, pinnatipeds or *pinnatipedes* and palmipeds or *palmipedes*).

After completing his studies at Oxford, Thomas Pennant (1726-1798) undertook several trips to Europe and visited Iceland in 1754.

The Gyr Falcon (*Falco rusticolus*) drawn by Petit Pailleu, plate taken from The British Zoology by Thomas Pennant (1726-1798).

At 81 years of age, Latham began to publish a new work on birds, *General History of Birds* (1821–1828), in which he described numerous new species studied in private collections. Unfortunately, since these have disappeared, we do not know which species to ascribe Latham's descriptions to. Paradoxically, he cites the binomials of the species already described, but does not propose his own binomials for the new ones.

As this was the era of the intensive colonization of Australia, he studied a number of birds coming from that country, which is why he is sometimes considered to be the father of Australian ornithology despite never having left Great Britain. Australia would have to wait for the visit of John Gould and the publication of his works (see page 169) for a systematic description of its fauna.

The final quarter of the 18th century was a favourable period for English ornithology, with the publication of a number of books on birds: *Outlines of the Natural History of Great Britain* (1769) by John Berkenhout (1726–1791), *Ornithologia Britannica* (1771) by Marmaduke Tunstall (see

John Latham (1740-1837), christened the father of Australian ornithology, despite never leaving England, studied a great number of specimens sent back by explorers.

page 62), *A Natural History of British Birds* (1775) by illustrator William Hayes (active from 1780 to 1800), *Synopsis of British Birds* (1789) by John Walcott (1754–1831), *Birds of Great Britain* (1789–1794) by William Lewin (1747–1795), *Entire New System of Ornithology: or Aecumenical History of British Birds* (1791) by Thomas Lord (*?–c.*1796) and Dr Duprée and *Natural History of British Birds* (1794–1819) by Edward Donovan (1768–1837).

The other classifiers of birds: Barrère, Klein, Möhring and Schäffer

The 18th century is characterized by a love of classification. This was very much the case in ornithology in the second half of the century. Several attempts at classification would be made, but none would achieve the success of Linnaeus and they would have a lesser impact on the history of ornithology. Physician **Pierre Barrère** (*c.*1690–1755) proposed, in his *Ornithologiae specimen novum… in classes, genera et species, nova methoda, digesta*, published at Perpiganan in 1745, the classification of birds based solely on the shape of their feet. This physician, who spent five years in French Guiana, grouped them into palmipeds, semipalmipeds, fissipeds and semifissipeds. Within each of these four groups, species were classified in a somewhat random fashion, mixing large and small birds (it is highly likely that this was an attempt to simplify their arrangement in the cabinets of curiosities). Unlike the work of Ray and Willughby, his classification was completely artificial, relying on a single anatomical criterion. Barrère arranged closely related birds living in the same habitats into different families.

Many other naturalists followed the Linnaean system in part, but ended up with quite different classifications. **Jacob Theodor Klein** (1685–1759), a German civil servant and amateur naturalist, assembled a large cabinet of curiosities and published a number of books. He sought to create a system of classification for the animal kingdom, but it would be quickly supplanted by that of Linnaeus. In 1750, he released *Historiae avium prodromus* in which birds are classified according to the shape of their beaks and feet.

Paul Heinrich Gerhard Möhring (1710–1792), German physician to one of the Princes of Anhalt, grouped birds into four classes in his book *Avium genera* (1752). The first, Hymenopodes, is divided into Piciae, which brought together woodpeckers and other piciformes plus a number of passerines, and Passeres, comprising the other passerines. The second class, Dermatopodes, is divided into Accipitres, which includes

ORNITHOLOGIÆ
SPECIMEN NOVUM,

SIVE

SERIES AVIUM IN RUSCINONE,
Pyrenæis Montibus, atque in Gallia Aequi-
noctiali Observatarum, in Classes, genera
& species, nova methodo, digesta.

*Auctore PETRO BARRERE, Societatis
Regiæ Scientiarum Monspeliensis Socio, in Acade-
miâ Perpinianensi Medicinæ Professore, Nosoco-
mii Regii Militum Mediæve, in Insulâ Gallo-Ame-
ricanâ Cayennâ, olim Medico ac Botanico Regio.*

PERPINIANI,
Apud GUILL. SIMONEM LE C.
Regis Typographum.

M. DCC. XLV.

Frontispiece (above) and plate (below) from *Ornithologia specimen novum* (1745) by Pierre Barrère.

birds of prey, parrots and some nightjars, and Gallinae, which grouped together the gallinaceans and pigeons. The third, Brachypterae, included ratites, bustards and the dodo. The forth, Hydrophilae, was subdivided into five orders: Odontorhynchae (flamingos, ducks, anhingas), Platyrhyncae (penguins), Stenorhynchae (pelicans, cormorants, tropicbirds, gulls, skuas), Urinatrices (divers and coots), Scolopaces (cranes, ibises, herons, rails, storks, guinea fowl, hummingbirds, waders). He completed his classification with a list of unclassifiable birds.

German naturalist **Jacob Christian Schäffer** (1718–1790), in possession of an impressive cabinet of curiosities, published *Elementa Ornithologica* in 1774, in which he also used the shape of the feet to classify birds into two groups, Nudipedes and Plumipedes, corresponding to aquatic and terrestrial birds. Schäffer subdivided these two groups into several subgroups, also based on the shape of their feet.

The publication of the works of Georges Cuvier (see page 183) at the end of the century would mark a change in the way classifications were approached, no longer being based on isolated anatomical criteria. They represented a shift towards thinking about the organization of living organisms as a whole.

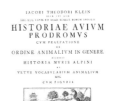

Frontispiece of *Historiae avium prodromus* (1750) by Jacob Theodor Klein.

Jacob Christian Schäffer (1718-1790) was interested in many fields, though primarily entomology.

The popular appeal of natural history, a new interest in regional fauna

An interest in flora and fauna was not solely the preserve of wealthy owners of large collections or daring explorers. More and more amateurs devoted their leisure time, and sometimes their entire lives, to the study of nature. Not having access to exotic specimens, they studied the fauna around them. Little by little towards the end of the century and the beginning of the next, a whole literature describing the fauna of different regions of Europe developed. A veritable infatuation with the study of nature emerged, encouraged by the Linnaean system.

In the United Kingdom, Reverend **Gilbert White** (1720–1793), a pastor who travelled all over his parish of Selborne, is one example. Each region, each kingdom, he wrote, must possess its own monograph on birds, and it is this type of work that is in the best position to aid the progression of natural history. He affirmed the importance of direct observations in building lists of species, as well as for better understanding migration and other behaviour. In 1767, he met Thomas Pennant (see page 82) with whom he would remain in correspondence. This correspondence would serve as the basis for a popular book, *The Natural History and Antiquities of*

Cornelis Nozeman (1721-1786) published the first description of the birds of the Netherlands in 1770. Plate by Jan Christiaan Sepp (1739-1811), taken from *Nederlandsche Vogelen,* depicting a Common Shelduck (*Tadorna tadorna*).

Selborne, published in 1789. He proved to be a fine observer of the neighbouring countryside even if he did not manage to solve the mystery of the supposed wintering of swallows in chimneys. He discussed nest building, migrations, feeding habits and seasonal plumage changes with his contacts. His book had a considerable impact and would see over 150 different editions right up until the present day. White corresponded with Buffon, who used his observations in his own works.

A History of British Birds, published in 1797 and 1804 by **Thomas Bewick** (1753–1828) had a more ambitious range than that of White's correspondence. In it he described all birds known in Great Britain. He became an apprentice of Ralph Beilby (1743–1817), an engraver from Newcastle, at 14 years of age. His first engravings on wood appeared in *Mensuration* (1768), which was the work of mathematician

Charles Hutton (1737-1823). Remaining in collaboration with Beilby, in 1785 he began to prepare the woodcuts for *General History of British Quadrupeds* that would be published five years later. The work was reprinted many times in subsequent years, until the eighth edition in 1804.

Bewick began to work on the illustrations of his work on birds. His partnership with Beilby was interrupted in 1797, the year of the release of the first volume of *A History of British Birds*, devoted to terrestrial birds. It was followed, in 1804, by a second volume devoted to aquatic birds.

His relatively inexpensive books were very successful, thanks largely to the excellent quality of the 233 woodcuts. These would be reused in a large number of works, up until the 20th century. The texts drew largely from the work of Thomas Pennant (see page 82) as well as that of other naturalists. Even if he did not bring to light new knowledge on birds, Bewick contributed to popularizing the study of birds and influenced a number of young people who would become ornithologists.

His contribution to ornithology was honoured by John James Audubon (see page 163) who came to visit Bewick in 1827 and dedicated the Bewick's Wren (*Thryomanes bewickii*) to him in the same year. William Yarrell (1784-1856) would name Bewick's Swan (*Cygnus bewickii*) after him in 1830.

In France, the situation was quite different. It would not be until the beginning of the next century that books on regional fauna would begin to appear with the publication of *Tables méthodiques des mammifères et des oiseaux observés dans le département de la Haute-Garonne* (1798-1799) by Philippe

Thomas Bewick (1753–1828), an engraver who became a naturalist.

Thomas Bewick's woodcuts are instantly recognizable. They would be reused in a large number of works.

The sheer number of popular illustrated works was testament to the popularity of ornithology in the 18th century. Maddalena Bouchard was inspired by the illustrations in the works of ornithologists Xaverio Manetti (1723-1784) and Franco Andrea Bonelli (1784-1830) in creating *Recueil de cent-trente-trois oiseaux des plus belle* [sic] *espèces* (1775), to which she added some personal observations.

Le Grand Duc de Virginie. La Hulotte. Strix maior sive Strix Aluco. Strige maggiore, o Strige Alloco.

Isidore Picot de Lapeyrouse (1744-1818) and *Ornithologie provençale* (its publication began in 1825, but would remain unfinished) by Polydore Roux (see page 123).

In Germany, **Johann Matthäus Bechstein** (1757-1822) did not have access to any collections. To complete his books, *Gemeinnützige Naturgeschichte Deutschlands nach allen drey Reichen* (1789-1795), of which three of the four volumes are devoted to birds, and *Ornithologisches Taschenbuch* (1802-1812). Bechstein explored his native Thuringia, corresponded with other naturalists and made use of the publications of his period. His direct observations, through which he discovered many new species, would prefigure those of Johann Andreas Naumann (1744-1826) and his son Johann Friedrich Naumann (1780-1857).

In Italy, **Francesco Cetti** (1726-1778) contributed to knowledge of the avifauna of Sardinia. Born in the south of Germany of parents of Lombardic origin, he joined the church in 1760 and became a member of the Society of Jesus. He responded to the call of the King of Sardinia, Charles Emmanuel III, who sought to modernize teaching on the island by entrusting it to Jesuits. In 1765, Cetti received the mathematics chair of the University of Sassari. He studied the surrounding fauna and flora and published *Storia naturale della Sardegna* in 1774, 1776 and 1777. His works, particularly those on the mammals and birds of the island, were of high quality and would long remain unequalled. The bird that bears his name, Cetti's Warbler (*Cettia cetti*), would only be discovered 40 years later by Alberto della Marmora (1789–1863).

The transit of Venus, or the multiplication of great expeditions

In the middle of the century, an astronomical phenomenon would garner considerable scientific interest: the double transit of Venus. This phenomenon, which only occurs every 120 years, consists of two passes of Venus in front of the sun in a eight year interval. The first, in 1761, allowed for several astronomical discoveries which demonstrated the presence of an atmosphere around Venus. To observe the second occurence in 1769, the Western powers organized great voyages to Siberia, Madagascar, New Guinea, the Pacific and Norway. These missions allowed for the accurate calculation of the distance between the sun and earth. Naturalists accompanied the astronomers in these expeditions, and undertook a large number of observations. The works of Linnaeus, Brisson and Buffon listed approximately 1,500 species of birds, but this number, thanks to this new initiative, would rapidly increase.

● The discovery of Eurasia

Peter Simon Pallas (1741–1811) led one of these missions to Siberia. Pallas was one of the many Germans who were called by the Russian Imperial Court to work at the Academy of Saint Petersburg. The aim was a political one: to compete with the scientific power of the states of Western Europe, and also to explore Siberia and central Eurasia in order to assess and exploit its natural resources.

The son of the head surgeon of a hospital in Berlin, Pallas learned Latin, French and English from a very young age. A brilliant student, he received training in surgery, as well as botany and anatomy from his father. Pallas was particularly interested by animals and conceived his own system of classification for birds at age 15. He became a medical doctor at Leiden at 19 years of age, and wrote his thesis on intestinal parasites. He was one of the first zoologists to adhere to the Linnaean system, which earned him the recognition of the Swedish scholar. He studied mammals, including new species

Peter Simon Pallas (1741–1811), a silhouette by Johann Friedrich Anting (1735–1805) from 1784, a style of portrait that was very much in fashion at the time.

from the zoo kept by William V, Prince of Orange (1748–1806). He began to publish monographs about these from 1767. Pallas made use of anatomical dissections in establishing the species that he described, allowing him to make comparisons with other animals. In this way, he was much more exacting that Linnaeus and his students who were satisfied with the use of simple, unique criteria. Not in possession of a reputation and unable to find a position that suited him, he contemplated returning to Berlin to practise medicine, when in 1767, he received a letter from Empress Catherine the Great (1729–1796) offering him a position in the Academy of Sciences of Saint Petersburg.

In 1768, shortly after his arrival in Russia, Pallas set off for what would be six years of exploration of the still largely unknown areas of the Ural Mountains, Altai, Lake Baikal, the Caspian Sea and South China. Naturalist travellers employed by Catherine the Great had an obligation to write records of their observations during the winter and immediately send them to St. Petersburg, a rule created in the fear that the explorers would die before the end of their journey. It is in this way that the beginning of the record of Pallas's journey appeared in 1771 under the title *Reisen durch verschiedene Provinzen des Russischen Reiches*. Pallas described the fauna and flora, but also fossils and minerals of the areas he visited. He also took an interest in the peoples that he encountered and put together dictionaries of their languages. In this first volume and in spite of the absence of documentation, Pallas described six new species of birds. When he returned from his six years of exploration, he discovered that his name was already well known throughout Europe.

At Saint Petersburg, Pallas studied the collections assembled from different expeditions, including those of Daniel Gottlieb Messerschmidt (1685–1735) who travelled to Siberia from 1720 to 1727, and Georg Wilhelm Steller (1709–1746) who would marry the former's widow and explored the Kamchatka Penninsula from 1740 to 1746. Pallas played a very important role during this period – a fact that can partly be explained by his good relationship with the Empress, who confided in him. Having become one of the patrons of Russian science, Pallas also organized similar expeditions to the coastal regions between Russia and America from 1786 to 1794.

His primary works were *Flora Rossica* (1784–1788) and *Zoographia Rosso-Asiatica...* He spent 15 years working on the latter, but death would prevent him from ever seeing its publication. Unfortunately, following the bankruptcy of his engraver, C.G.H. Geissler, the work appeared without illustrations and the complete edition would not be available

Johann Anton Güldenstädt (1745–1781) supervised the publication of the record of P.S. Pallas's voyage.

Plate depicting White-headed Ducks taken from the account of the voyage of Peter Simon Pallas (1794).

until 1831, 20 years after Pallas's death. *Zoographia Rosso-Asiatica* would become a primary reference for researchers for several decades, but the theories developed by Pallas would exert even greater influence.

He carried out a synthesis of Linnaeus's binomial nomenclature and Buffon's ideas on the variability of species. Pallas was interested in geographic variations and, even if he did not specify them, he described them with great care. He should rightly be considered as one of the founders of biogeography. His ideas were rapidly taken up by others, including Constantin Wilhelm Lambert Gloger (1803–1863) in his small book *Schlesiens Wirbelthier-Fauna* (1833). In it, he described the effect of climate on the appearance of individuals and was the first to separate swallows from swifts. His ideas most certainly influenced Charles Darwin (1809–1882).

In the following century, Russia would continue to attract large numbers of Germans. Johann Fischer von Waldheim (1771–1853) taught at Moscow and devoted himself to the study of Siberian birds towards the end of his life, Friedrich Heinrich von Kittlitz (1799–1874) participated in the first Russian expedition to circumnavigate the globe, during which he studied the birds of Chile among others, Leopold von

Schrenck (1826–1894) travelled to the island of Sakhalin and the region of the Amur River, and Gustav Radde (1831–1903) devoted himself to the study of the birds of Caucasus. To these Germans, we can add Frenchman Édouard Ménétries (1802–1861) who participated in a Russian mission to Brazil and took over the management of the zoological collection of the Academy of Science in St Petersburg.

Though he was not one of these expatriates, **Blasius Merrem** (1761–1824) was profoundly influenced by Pallas. Merrem was one of the many ornithological students of Johann Friedrich Blumenbach (1752–1840) who taught anatomy during 60 years at the University of Göttingen and whose work on human races remains well known. Merrem did not have an easy life in a Germany troubled by the disturbances of the late 18th century. A man of modest origins, he was obliged to become a jurist to earn his living and also gave night courses to supplement his income. All this left him with little time for zoology, his true passion, and all of his projects would remain unfinished. Nonetheless, the classification of birds proposed by Merrem had a significant influence on the ornithologists of his time, being superior to those published before it. Some even consider it to be at the root of all subsequent work on classifications. He introduced the differentiation of ratites (flightless birds) and *Carinatae* (flying birds) based on the presence or absence of a keel on the sternum. Carrying on the work of Dutchman Peter Camper (1722–1789) and Englishman John Hunter (*c.*1754–1809) on bird bones, he also discovered the existence of pulmonary sacs.

Blasius Merrem (1761–1824), in spite of a difficult life, had a major influence on ornithological and herpetological classification. Like his employer, P.S. Pallas, he attempted to reconcile the Linnaean system with that of Buffon by according an important place to anatomy in classification.

• The great round-the-world journeys by sea

In parallel to the journeys over land, many circumnavigations were organized at the time of the transit of Venus. The botanist Sir Joseph Banks (1743–1820), a naturalist and student of Linnaeus named Daniel Solander (1733–1782) and the naturalist and illustrator Herman Diedrich Spöring (1733–1771) accompanied astronomer Charles Green (1735–1771) on a monumental mission to the Pacific that received the support of the Royal Society. The secret objective of this expedition was more prosaic: the British admiralty wanted to uncover the existence of a large austral continent that was supposed to exist but for which there was no proof. Commanded by James Cook (1728–1779), HMS *Endeavour* arrived at Tahiti in 1769. The mission was a success. The specimens brought back and the account of the expedition would garner an immense amount of interest throughout Europe (and the gentle life of the Tahitian archipelago would also fuel the pre-

Romantic imagination). As soon as he returned, after more than three years of travel, Cook was awarded a new mission and departed again in 1772. Banks refused to accompany him and he was replaced by Johann Reinhold Forster (1729–1798), his son Johann Georg Adam Forster (1754–1794) and another student of Linnaeus, Anders Sparrman (1748–1820). Cook's third voyage would also be his last. He would be killed by indigenous islanders during a stopover in Hawaii in 1779.

Still with the objective of observing the transit of Venus, Louis XV appointed Louis Antoine de Bougainville (1729–1811) to conduct the first French circumnavigation of the globe. This voyage, organized with the input of a number of prominent scientists, including astronomers like Maupertuis and Lalande, as well as Buffon, would only count one naturalist among its ranks, Philibert Commerson (1727–1773). These long voyages were not without their dangers and entire expeditions were known to disappear. Such was the fate of an expedition run by Jean-François de La Pérouse (1741–1788).

The bounty brought back from these voyages was exceptional: 28,000 natural specimens gathered during Cook's third voyage. This wealth of specimens was shared between the

A Southern Fulmar (*Fulmarus glacialis*) painted by Georg Forster during his round the world journey.

Plate of an Australian Gang-gang Cockatoo (*Callocephalon fimbriatum*) taken from the atlas that accompanied *Voyage autour du monde par la frégate du roi la* Boudeuse *et la flûte* l'Étoile, *en 1766, 1767, 1768 1769* by Louis Antoine de Bougainville.

collection of Sir Joseph Banks and Sir Ashton Lever (see page 62). The observations derived from these collections were often imperfect: some of the specimens gathered by Commerson were lost and his notes, used by many scholars, including Buffon, were never fully exploited.

The birth of American ornithology: William Bartram

The British colonies of the New World claimed their independence in 1776, and a new nation was born: the United States of America. This country would become one of the pre-eminent centres for science in the world.

William Bartram (1739–1823) was one of the first American naturalists. His love of nature was shared by his father, John Bartram (1699–1777), the great botanist and founder of a botanic garden in Pennsylvania, the first of its kind in the United States. At 16 years of age, William Bartram sent bird skins and drawings to George Edwards (see page 66). At around 30 years of age, he tried his hand at rice growing in Florida, but without success. He returned to Pennsylvania where he took up a number of trades without much luck. In 1791, he published the record of his five-year experience in Florida: *Travels through North and South Carolina, Georgia, East and West Florida, the Cherokee Country, the Extensive Territories of the Muscogulges or Creek Confederacy, and the Country of the Chactaws.* This book would not achieve any measure of success in America, but would see nine editions in Europe, fuelling the Romantic British imagination.

Bartram would continue to collect plants and observe American birds for the rest of his life. His list would consist of 215 species. He played an important role in the development of American natural history. At 60 years of age, he met Alexander

Portrait of William Bartram (1739-1823) painted by Charles Willson Peale.

Purple Finch (*Carpodacus purpureus*) with two plants (*Illicium floridanum* and *Houstonia serpyllifolia*), plate by William Bartram.

Wilson (see page 127), then a schoolteacher, and introduced him to ornithology, sharing his experience and encouragement. He had a profound impact on his great-nephew, Thomas Say (1787–1834), who often came to help him with the work in his botanical garden. Wilson would become a founder of American ornithology, while Say would fulfil that role in entomology. Réné Primevère Lesson dedicated the genus *Bartramia* (of which Upland Sandpiper *Bartramia longicauda* is the only member) to William Bartram in 1831.

The city of Philadelphia would play a central role in the development of science in the newly formed America. It was in this city that the first private museum was opened in 1784 by **Charles Willson Peale** (1741–1827). Besides his interest in natural history, Peale was a renowned painter and some of his many children (whose first names included Rubens, Titian, Rembrandt and Raphaelle) were artists. Charles Willson Peale's museum in Philadelphia contained no fewer than 760 species in 140 cases. The classification of birds in the museum followed the Linnaean system. This collection contained various birds from Alexander Wilson, Thomas Say and Charles Bonaparte. Peale bequeathed his collection to one of his sons, Charles Linnaeus, but the latter would have the collection auctioned off in 1845. A number of specimens were acquired by Phineas Taylor Barnum (1810–1891), the famous showman entrepreneur, but would be lost in a storage facility fire in New York.

The 18th century ended with the tumult of the French Revolution and the Napoleonic wars. Following these a long period of prosperity for the West ensued: the colonisation of the world showed renewed vigour, and scientific and technical progress accelerated. Thus began the golden age of ornithology.

Charles Willson Peale (1741–1827) unveiling his collection of birds.

The 19th century
The golden age of ornithology

I n the 19th century, everyone seemed to have a passion for nature and collecting, fuelling demand for a large market in natural specimens. The scholars of the West gathered in learned societies and it was a good time to be a botanist, entomologist, conchologist or ornithologist. Books and journals, often very luxurious, would gain record popularity.

This interest was sustained by technical innovations (like taxidermy, which simplified the preservation of specimens), by expeditions that gradually became faster and easier to carry out, and more generally by improvements in the standard of living in Western societies.

Exploration of the world gave way to exploitation. The great contemporary colonial empires took their places on the world stage. This colonization, supported by large populations of civil servants and troops, contributed greatly to the discovery of overseas fauna. The young American nation would quickly rise to prominence and take its place amongst the first ranks of scientific nations.

The 19th century was also an era in which the profound disruptions to nature began to be felt, accompanied by the nostalgia of an increasingly urbanized populace. Little by little, the imperative to protect habitats and species would come to the fore.

Summary

Scientific expeditions: how ornithologists discovered the world

As we saw in the preceding chapter, expeditions often incorporated a scientific component which included the study of the fauna and the flora of the regions explored. The first educational pathways for naturalist-voyagers opened up in 1758, leading to the publication of works such as *Mémoire instructif sur la manière de rassembler, de préparer, de conserver et d'envoyer les diverses curiosités d'histoire naturelle*. In addition to the proper preservation of specimens and seeds, it sought to draw attention to assessing the value of objects.

The research institutions that emerged after the French Revolution played active role in the writing of these texts. The Muséum de Paris, founded in 1793 from the Jardin du Roi, trained its own naturalist-voyagers who would set out for many different destination, including South Africa, South America and Australia. This was not an isolated example. The museum in Vienna sent Johann Natterer (1787–1843) to Brazil, the government of Prussia sent Friedrich Sellow (1789–1831) to South America and the Netherlands put Caspar Georg Carl Reinwardt (1773–1854) in charge of founding the botanical garden in Buitenzorg, which supplied specimens to the Rijksmuseum van Natuurlijke Historie in Leiden.

It is in this context that the Dutch government organized a commission with the specific task of studying the fauna and flora of its colonies, the **Natuurkundige Commissie**, operating from 1820 to 1850. It sent young naturalists, all medical graduates, to the Dutch possessions to gather specimens. The living conditions were tough and most would die prematurely, sometimes only a few months after their arrival: Christiaan Macklot (1799–1832) at age 33, Eltio Alegondas Forsten (1811–1843) at age 32, Heinrich Kuhl (1797–1821) at age 24, Johan Coenraad van Hasselt (1797–1823) at age 26 and Heinrich Boie (1794–1827) at 31 years old. Only three naturalists would survive their mission, among them Salomon Müller (1804–1864), who gathered 6,500 bird skins, as well as skeletons, nests and eggs.

The pursuit of life sciences represented significant political stakes for the leading nations of the time. Napoleon did not hesitate to requisition the richest cabinets of curiosities from the countries he entered with his army. This resulted in the transferring of the collection of the Prince of Orange in The Hague to Paris in 1795. In 1808, a delegation of scholars, led by Étienne Geoffroy Saint-Hilaire (1772–1844), made its way to Lisbon, then occupied by French troops, with the specific

Plate taken from *Description de l'Égypte* (1820), a vast work undertaken by the scientists of the *Commission des sciences et arts d'Égypte* established by Napoleon.

objective of bringing back the most interesting Portuguese collections. Napoleon's taste for scientific knowledge and his desire to grow the scientific influence of France would result in the organization of many scientific missions. His military expedition to Egypt was joined by the *Commission des sciences et arts d'Égypte* (including 10 or so naturalists), which would contribute to advancing the knowledge of the avifauna of the Middle East. It was Victor Audouin (1797-1841), author of famous contributions to entomology, who studied the specimens brought back from the expedition. He published his findings in *Explication sommaire des planches d'oiseaux de l'Égypte et de la Syrie* in 1826.

Expeditions were not only commissioned by the ruling powers of the day, they were very often undertaken by private naturalists. These expeditions represented, for many young scientists, the opportunity to make a name for themselves, should the collections they returned with prove interesting and novel. Their study would make for numerous publications and attract the attention of chief naturalists, in the hopes of having their efforts rewarded, notably by being offered a permanent position. These private expeditions were financed by the sale of specimens to collectors and institutions in Europe. Such was the case of botanist William John Burchell (1781-1863) who financed his travels to South America and South Africa through the sale of the zoological specimens he collected.

We will now turn to some of the striking and more typical examples of the ornithological figures who chose to travel great distances, risking their lives for their work.

- François Levaillant: traveller and inveterate liar?

The unusual story of the birth of **François Levaillant** (1753-1824) has at times been called into question: his father, Étienne François Levaillant, not having been able to secure his future wife's hand in marriage, is said to have run off with her. The couple fled to South America in 1751 and settled at Paramaribo, a Dutch colony belonging to the Dutch East India Company. Étienne Levaillant became a businessman, and was appointed to represent the interests of France in the region. François Levaillant would reminisce on his childhood in America in the introduction of *Voyage de Monsieur de Le Vaillant dans l'intérieur de l'Afrique* (1790). He explained that his parents, who were passionate about natural history, frequently took him into the jungle to collect specimens. Having mastered the use of a blowpipe, he began to build his own cabinet of curiosities at a young age by hunting the birds in the forest. On returning to Europe at the age of 10, Levaillant studied in

François Levaillant (1753-1824), who, despite the errors he committed and the alleged excesses of his imagination, was the author of the best works on birds from the early 19th century.

Germany and Alsace. It was at Metz that he learned the technique for preserving birds, as well as the art of hunting and observing them from Jean-Baptiste Bécoeur (1718–1777). Bécoeur possessed one of the richest collections of the period, with more than 2,000 specimens, far surpassing that of the cabinet of the King in Paris, which contained only 463.

From 1777 to 1780, Levaillant stayed in Paris and devoted himself to the study of the birds preserved in the rich collections of the capital, particularly that of Pierre Jean Claude Mauduyt de la Varenne (c.1732–1792). Bécoeur and Mauduyt de la Varenne's collections, acquired by Duke Charles II of Pfalz-Zweibrücken, would be completely destroyed in the fire ignited by French troops at the ducal palace in 1793.

However, as he wrote himself, "these superb displays soon caused me much grief, they left my soul empty in a way that nothing else could". He then decided to travel and gained the support of the influential Jacob Temminck, treasurer of the Dutch East India Company, which possessed one of the best private collections of birds in Europe. The company would send him to South Africa, an area that Levaillant explored for nearly five years. He returned to Paris in 1784, with more than 2,000 bird skins, complaining bitterly of the indifference with which he was received on his return in his writings. He then attempted to sell his collection to the Muséum d'Histoire Naturelle in Paris at a reduced rate, but was not successful. It was finally Temminck who would buy it back in part. It is now part of the collections of Museum of Leiden in the Netherlands. Briefly imprisoned during the French Revolution, Levaillant regained his freedom with the fall of Robespierre in 1794.

The illustrations in his works were of a high calibre. A number of them were signed by Jacques Barraband (1768–1809), a student at the Ecole des Gobelins who worked in the manufacturing of china. *Histoire naturelle des perroquets* would be completed in 1837 by Alexandre Bourjot Saint-Hilaire (1801–1886).

In 1796, Levaillant published a book in which he provided the account of his travels, *Second voyage dans l'intérieur de l'Afrique,* but he is best known for the many works that he devoted to birds including *Histoire naturelle des oiseaux d'Afrique* (1796–1808), *Histoire naturelle d'une partie d'oiseaux nouveaux et rares de l'Amérique et des Indes* (1801), *Histoire naturelle des oiseaux de paradis (1801–1806), Histoire naturelle des perroquets* (1801–1805), *Histoire naturelle des cotingas et des todiers* (1804) and *Histoire naturelle des calaos* (1804). While the first books were extremely successful, the later ones sold poorly. He had a profound influence on a number of ornithologists as C.J. Temminck testified (see page 142) in 1825: "I owe to Levaillant

Guianan Cock-of-the-rock (*Rupicola rupicola*) engraving by Barraband, taken from *Histoire naturelle des oiseaux de paradis et des rolliers, suivie de celle des toucans et des barbus* (1806) by François Levaillant.

and to his writings my first thoughts on and my first foray into natural history."

Following Buffon's lead, he opposed the Linnaean system, which he thought too artificial. For him, the best way to differentiate species was to study them in nature, the only way that species could be situated in the natural order. He also believed that "any description written based on a single individual that was seen by chance, will always be imperfect and insufficient for correctly recognizing a species, because it is only after multiple comparisons and reiterations in different seasons, on a great number of birds of the same species that it is possible to attach any degree of certainty to it ... a knowledge without which the history of birds will always be confused and full of errors".

Levaillant, rejecting Linnaeus's binomial system, preferred to use often very simple names for species, such as the 'Singer', the 'Clawed', the 'Huppard', the 'Mocker' or the 'Vociferous'. He also enjoyed onomatopoeia, naming species for their calls,

Illustration of a Bird of Paradise by Barraband taken from *l'Histoire naturelle des oiseaux de paradis et des rolliers, suivie de celle des toucans et des barbus* (1806) by François Levaillant.

for example the Brubru or the boubou. Moreover, Levaillant gave no indication of their classification. The species he described for the first time received their scientific names from the supporters of the Linnaean system like F.M. Daudin (1774–1804), Louis Vieillot (1748–1831) or Dumont de Sainte-Croix (1758–1830).

In 1857, Carl Jakob Sundevall (1801–1875) undertook a very critical analysis of *Histoire naturelle des oiseaux d'Afrique* on account of the fact that some of the species described by Levaillant could not be found 60 years later. Of 284 species described in the work, 10 are not identifiable, 71 were obtained from outside Africa, 50 are doubtful and 10 are considered to have been falsified from the collected feathers of multiple birds. It is also suspected that Levaillant had invented a portion of his peregrination. It was not until the rediscovery of hitherto unknown illustrations in the library of the South African parliament and through the careful study of the locations described by Levaillant that the truth about his travels in the region of Orange River and Namaqualand would finally come out. However, some of the species that were not recognizable may in fact have since disappeared. The numerous errors and inventions of the Frenchman remain, making him somewhat of an enigma.

- German travellers explore Africa

The absence of large ornithological collections in Germany in the 18th century certainly explains this country's lagging behind in descriptive ornithology. Once the Napoleonic threat had passed, Germany would quickly catch up and become one of of the world's foremost centres for ornithology. The appearance of numerous museums throughout the country from 1825 echoed this enthusiasm: the Senckenberg Museum in Frankfurt am Main directed by Philipp Jakob Cretzschmar (1786-1845), the museum in Darmstadt directed by Jakob Kaup (1803-1873), the museum in Munich directed by Johann Baptist von Spix (1781-1826) and that of Halle directed by Christian Ludwig Nitzsch (1782-1837). To these, we should also add the private museum of Prince Maximilian of Wied-Neuwied (1782-1867), famed for his expedition to Brazil and then to North America. These institutions had many beneficial side effects, notably providing stable positions for scientists and building up reference collections. The desire for specimens was behind the financing of many expeditions, mainly in Africa.

North and East Africa attracted many generations of German explorers. **Friedrich Wilhelm Hemprich** (1796-1825) and his friend **Christian Gottfried Ehrenberg** (1795-1876) joined an expedition financed by the Academy of Berlin to explore Egypt and Abyssinia between 1821 and 1825. The expedition was a success, leading to the collection of 46,000 botanical specimens and nearly 34,000 animal specimens, including numerous species of birds. **Eduard Rüppell** (1794-1884) and **Michael Hey** (1798-1832) were acquainted thanks to the initiative of P.J. Cretzschmar. They explored North and East Africa over several long expeditions during the first third of the century (the collections that they assembled are conserved at the Senckenberg Museum in Frankfurt). The section of the expedition account devoted to ornithology described about 30 new species, among them Meyer's Parrot (*Poicephalus meyeri*), Nubian Bustard (*Neotis nuba*), Goliath Heron (*Ardea goliath*), and Streaked Scrub-warbler (*Scotocerca inquieta*), all of which were described by P.J. Cretzschmar.

Eduard Rüppell (1794-1884) was the first scientist to study the avifauna of Abyssinia.

Theodor von Heuglin (1824-1876) continued the exploration undertaken in the African regions and explored the east of the continent during several expeditions between 1852 and 1875. He published two volumes of *Ornithologie Nord-Ost*

Afrikas (1869–1874), which became a reference work on the avifauna of this part of the continent.

Theobald Johannes Krüper (1829–1921), after having completed his thesis on the distribution of falcons in Europe, travelled extensively: in Lapland in 1855, Iceland in 1856 and Sweden in 1857. His subsequent explorations would take him to Greece and Turkey, regions that had been largely ignored by naturalists up until this point. He assumed the directorship of the museum of the University of Athens in 1872, but his position was so poorly paid that he had to continue to sell specimens to European museums and collectors. His pioneering work would encourage many specialists to take an interest in the fauna of these regions.

- The explorers of the French Navy

Naval physicians would play a significant role in the scientific explorations of the world throughout this century. Each ship setting out for long distance travel required a physician. Medicine was for a long time the only form of scientific training, which meant that in addition to their medical role, they were entrusted with the task of collecting biological specimens and ensuring their conservation.

In France, the disappearance of the expedition led by La Pérouse in 1788 and the troubles of the French Revolution disrupted the organization of expeditions for some time. Expeditions would start again in 1800, with the launching of *Le Géographe* and *Le Naturaliste*, two vessels that sought to establish a permanent outpost in the southern seas before the British. In addition to some important cartographic surveys, a zoological collection of more than 100,000 specimens was assembled, containing nearly 2,500 new species. However, the journey was a rough one: its captain Nicolas Baudin (1754–1803) and two astronomers died and several naturalists, frightened by the rigours of the expedition, deserted it. It was Vieillot who described many species of the birds collected in the *Nouveau dictionnaire d'histoire naturelle*.

René Primevère Lesson (1794–1849) undertook one of the most important round-the-world voyages. This naval pharmacist was put in charge of the botanical garden of Rochefort, the seat of the naval medical school. He participated in a circumnavigation in 1822 aboard *La Coquille* that brought back a rich collection including 3,000 plants. The record of this voyage would not be published until 1838, as *Voyage autour du monde, entrepris par ordre du gouvernement sur la corvette La Coquille*. He received the Legion of Honour in 1825 and became a member of the Académie de Médecine three years later. In

René Primevère Lesson (1794-1849) devoted himself to ornithology after his round-the-world voyage.

The Ruby-throated Hummingbird (*Archilochus colubris*), plate engraved by Jean-Gabriel Prêtre, from *Trochilidées* by René Primevère Lesson.

1833, he would join the Académie des Sciences.

Lesson devoted himself to zoology and published many works on birds, including *Manuel d'ornithologie* in 1828 and *Traité d'ornithologie* in 1831. He was particularly interested in the birds of paradise and hummingbirds he had discovered during his travels. He was the first to describe a third of the species of hummingbirds. He regretfully noted in the introduction of his *Traité d'ornithologie* that "the study of birds does not benefit from the same amount of respect accorded to other branches of natural history in France". After having cited several works published in France, he continued: "Perhaps it is because of these expensive works that this science seems to be the exclusive domain of opulent amateurs, while at the same time the difficulties in bringing together and conserving specimens dissuades people who live far from the large public

Title page of Joseph Paul Gaimard's account of his expedition to Iceland and Greenland between 1835 and 1836.

collections from taking up their study." His classification would have no discernable impact. Lesson was also the author of other zoological and reference works, such as *Manuel d'histoire naturelle médicale et de pharmacographie* (1833).

Two other naval surgeons, **Joseph Paul Gaimard** (1796-1858) and **Jean René Constant Quoy** (1790-1869), were the authors of a great number of taxa. They participated, together and separately, in numerous scientific expeditions to Western Australia, Timor, the Moluccas, Samoa and Hawaii (1817-1820), Australia, New Zealand, Fiji, and Loyalty Islands (1826-1829), and Iceland and Greenland (1835-1836). The very large number of specimens brought back from these voyages enriched the collection of the Muséum National in Paris. The names of many birds have been dedicated in their honour, including the Red-legged Cormorant (*Phalacrocorax gaimardi*) by R.P. Lesson and Prosper Garnot (1794-1838), the Forest Eleania (*Myiopagis gaimardii*) by Alcide Dessalines d'Orbigny (1802-1857) and the Black Butcherbird (*Cracticus quoyi*) by R.P. Lesson.

- Otto Finsch, traveller, scientist and Prussian colonialist

Friedrich Hermann Otto Finsch (1839-1917) was typical of scientists who split their time between exploratory voyages and museum work. He was born in Silesia and his father hoped he would join the family business producing and marketing decorative glass. In 1858, having completed his apprenticeship, Finsch left for Bulgaria, where he was employed as a private tutor for a year. Showing a keen interest in natural history, he educated himself and at the age of 20 published his first article on the birds he observed in Bulgaria in the *Journal für Ornithologie*. He became an assistant at the museum in Leiden, where his new employer, Hermann Schlegel (see page 144), would introduce him to the techniques of conservation and complete his training. In 1864, Otto Finsch took the position of curator at the museum in Bremen, a position he would occupy for 14 years. His work on parrots, *Die Papageien*, published in two volumes, would be recognized with an honorary doctorate from the University of Bonn. From then on, he took an interest in the birds of the world. In 1867, with Gustav Hartlaub (1814-1900), he published a study on the avifauna of Polynesia, *Beiträge zur Fauna Central-Polynesiens*, and in 1870, he participated in the *Die Vögel Ost-Afrikas* study based on the collection of East African birds collected by Baron Karl Klaus von der Decken (1833-1865) during his expeditions.

Based on his study of the specimens at the museum in Bremen, he published a list of species on the avifauna of

Friedrich Hermann Otto Finsch (1839-1917) was a great naturalist and voyager. His most famous works dealt with parrots and East African and Polynesian birds.

Australia, New Zealand and Samoa. From the beginning of the 1870s, he began to contribute to *The Ibis* and visited the collections of the British Museum in London. The contact he made with British ornithologists would result in his election as an honorary member of the British Ornithologists' Union at the relatively young age of 32.

He then embarked on what would be numerous voyages. In 1871, he left for California, then, two years later, for Lapland. In 1876, he undertook an expedition of zoological study in western Siberia, accompanied by Alfred Edmund Brehm (1829–1884) and Count Karl von Waldburg-Zeil Trauchburg (1841–1890). This expedition would yield specimens of 283 species of birds.

In 1878, Finsch resigned from his position of conservator to devote himself to expeditions. With his wife Josephine, he visited Polynesia, New Zealand, Australia and New Guinea, where he could finally observe live examples of the birds he had only known through their skins. He arrived in America where he met Spencer Fullerton Baird (see page 134), director of the museum in Washington. From there, he left for San Francisco, Honolulu and the Marshall Islands. He remained several months in New Britain, a large island east of New Guinea. Finsch gathered more than 100 species and fared better than Theodor Kleinschmidt (1834–1881), a German naturalist who was killed by local people in the same area.

Finsch returned to Germany in November 1882 following another extended stay in Australia and New Zealand. In addition to his scientific objective, during his time in the east of New Guinea, a territory virtually unexplored by Westerners, he gathered useful information for Otto von Bismarck (1815–1898). This would culminate in the German colonization of much of north-eastern New Guinea and a series of islands including those of Buka and Bougainville, the Carolines and the Marianas (excluding Guam). The administrative seat was named Finschhafen in homage to Otto Finsch.

Finsch quickly realized that he could not rely on the support of politicians and, for a number of years, his earnings were limited to what he could earn from the written accounts of his travels. In 1889, he took the position of curator at the museum in Leiden. In 1895, he decided to return to Germany to take up the modest direction of the department of ethnography at the municipal museum in Brunswick, giving up ornithology to devote himself to ethnology. Following his unexpected death in January 1917, Germany would abandon its colonial possessions, including those acquired as a result of Finsch's expeditions.

Otto Finsch's contribution to the ornithological literature was considerable. He was the author of 14 works describing

Blue-and-yellow Tanager
(*Thraupis bonariensis*),
illustration by John Gould
from *The Zoology of the Voyage
of* HMS Beagle.

new genera and described about 150 new species. Many ornithologists honoured his work by dedicating species, such as the Lilac-crowned Parrot (*Amazona finschi*) dedicated by P.L. Sclater, the Grey-headed Parakeet (*Psittacula finschii*) by A.O. Hume, Finsch's Imperial-pigeon (*Ducula finschii*) by E.P. Ramsay (1842–1916), Finsch's Flycatcher-thrush (*Neocossyphus finschii*) by R.B. Sharpe and Finsch's Wheatear (*Oenanthe finschii*) by T. von Heuglin (1824–1876).

Scientific voyages would continue for the remainder of the 19th century. Little by little, the nature of the expeditions undertaken would change, as most regions of the world became known. For a long time *terra incognita*, Africa was the last continent to be fully explored. This delay can be explained by the ravages caused by yellow fever and malaria (30–70 per cent of settlers in tropical and equatorial Africa died during the first year of settlement). This obstacle was removed with the discovery of quinine, the first effective remedy against malaria. The world was changing. The aim was no longer to discover, but to colonize. The great colonial empires were taking shape, requiring hosts of civil servants and troops, and not forgetting

Charles Darwin's round the world voyage

One of the most famous voyages of the 19th century was undertaken by one of its most famous naturalists, Charles Darwin (1809–1882), who circumnavigated the globe aboard the HMS *Beagle* between 1831 and 1836.

Darwin's ornithological training was not very extensive. During his studies, he was primarily interested in plants and beetles. Nonetheless, he collected nests and eggs and learned taxidermy at the University of Edinburgh under ornithologist and explorer Charles Waterton (1782–1865). It was thanks to his enthusiasm for natural history and his mastery of Spanish (which he had learned following Humboldt's example and in the hope of exploring South America) that he was chosen for the expedition.

Darwin was given the responsibility of putting together the ship's library. The ornithological section consisted of scientific accounts of voyages (including those of Alcide Dessalines d'Orbigny and Maximilian zu Wied-Neuwied, J.B. von Spix and C.F.P. von Martius in South America, William John Burchell in Africa, Philip Parker King in Australia, and the circumnavigations of J.R. Forster, La Pérouse, Jean R.C. Quoy and J.P. Gaimard). It also included more general works such as *Dictionnaire classique d'histoire naturelle* by Bory de Saint-Vincent, *Manuel d'ornithologie* by R.P. Lesson, and Cuvier's *Le Règne animal* as well as its translation by Edward Griffith.

When he boarded the ship at 23 years of age, his understanding of ornithology both with respect to British and exotic fauna was rather weak. Nor was he in possession of reference collections and consequently was incapable of identifying a Short-eared Owl (*Asio flammeus*) when he captured one in the Falklands. He committed several errors in identifying birds, particularly marine birds. It was his young assistant, Syms Covington (1816–1861), who was responsible for hunting and preparing birds, Darwin preferring to devote himself to geological observations and the collection of plants and invertebrates.

From an ornithological point of view, his most important stop was certainly the one he made at the Galapagos islands, marked by the discovery of a group of species that would become known as Darwin's finches. Paradoxically, although Darwin was not an ornithologist, ornithology would play a large role in the development of his theory of evolution.

Although birds were not the central focus of his work, he brought back 468 bird skins from his five-year voyage, including ten specimens of Darwin's Rhea (*Pterocnemia pennata*), nests and eggs from 16 different species and roughly 20 specimens in alcohol. The 39 new species and subspecies, including the famous finches, were for the most part described in 1837 by John Gould (see page 169), ornithologist and member of the Zoological Society of London.

In spite of their historic significance, the birds brought back from the voyage were dispersed amongst at least eight different institutions and were not inventoried until 2004.

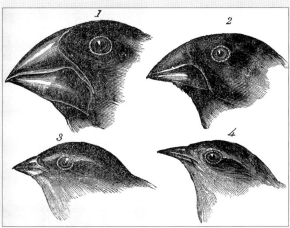

Four species of finch from the Galapagos (illustration taken from Darwin's *Journal of Researches*): 1. Large Ground-finch (*Geospiza magnirostris*) 2. Medium Ground-finch (*Geospiza fortis*) 3. Small Tree-finch (*Camarhynchus parvulus*) 4. Warbler Finch (*Certhidea olivacea*).

missionaries, among whose number there were many great ornithologists.

The British Empire and ornithology

The members of the colonial administration and army, like those employed by diplomatic bodies posted abroad, participated considerably to the understanding of local faunas, sending numerous specimens to European specialists. The English East India Company played a central role in this, so much so that it has been likened to a state within state.

• Colonial servicemen

Members of the military played a pivotal role in the formation of collections and other ornithological activities. Of the 650 members who joined the British Ornithologists' Union from 1858 to 1908, 86 were active or retired officers in the army. They were often important collectors and were much more efficient for having perfected their shooting skills. These servicemen devoted a portion of their leisure time to the hunt and often did not hesitate to collect specimens in the middle of a battle. Thus, Colonel Willoughby Verner (1852–1922) recounts, in the naturalist journal that he kept from 1867 to 1890, how he was able to collect a weaver's nest while under fire during a campaign in the Nile region. Several officers have greatly contributed to our understanding of Indian avifauna.

Major General Thomas Hardwicke (1755–1835) was stationed in India for more than 14 years, beginning in 1777. A serviceman in the East India Company, he had a passionate interest in natural history and was a pioneer in the exploration of India. Considering that it was easier to have the birds he hunted painted by local artists than to attempt to stuff them, he put together a collection of more than 4,500 paintings. He also remained in regular correspondence with scientists such as J. Banks of the Royal Society and J.E. Gray (see page 146) of the British Museum. The latter would make use of his illustrations in his *Illustrations of Indian Zoology* (1830–1834). Another pioneer was Colonel **William Henry Sykes** (1790–1872), posted in Bombay from 1805. He gathered an important collection that would be exhibited in the museum opened by the East India Company in London. Sykes was the first naturalist to attempt to create a catalogue of Indian birds and described no fewer than 56 new species in a publication on the birds of the Deccan Plateau in central India.

Major General Thomas Hardwicke (1755-1835) (above) and Colonel William Henry Sykes (1790-1872) were pioneers in the study of Indian birds.

Colonel **Samuel Richard Tickell** (1811–1875) described numerous species and varieties of birds and forwarded specimens to Blyth (see page 113) from Calcutta and to Hodgson (see page 114) from Kathmandu. An officer by training, he served several years in the infantry before joining the military administration. His duties in Bengal, and later Burma, allowed him to carry out many observations of the fauna of these regions. He retired in 1865, but suffered from an inflammatory attack a few years later that caused him to lose his sight. This would force him to interrupt a vast illustrated project on Indian birds. His illustrations, which depicted 276 species and were considered among the best of his time, have remained unpublished to this day. They are currently being conserved by the Zoological Society of London.

Sadly, Colonel Samuel Richard Tickell (1811–1875) lost his sight and would not have the opportunity to make use of his excellent observations.

- The physicians

As was the case with the great seafaring explorers, physicians played a prominent scientific role in this period. **Francis Buchanan-Hamilton** (1762–1829) completed his studies in medicine at the University of Edinburgh in 1783. One of his professors was none other than Sir James Edward Smith (1759–1828), who had purchased Carl Linnaeus's collections, correspondence and library from his widow. These objects were then transferred to London and their conservation assured through the creation of Linnaean Society of London in 1788.

Buchanan became a surgeon for the East India Company. His personal virtues and interest in natural history were noticed by Lord Richard Colley Wellesley (1760–1842), Governor General of India and an influential figure in the new British possessions. Wellesley conceived the ambitious project of listing all the mammals and birds of the Indian subcontinent, which would result 99 new species being described between 1801 and 1808. The menagerie created would serve as the beginnings of the Alipore Zoo.

Lord Wellesley entrusted Buchanan with a mission to explore and study the most southerly regions of India, before sending him to Kathmandu and Nepal. After a brief stay in England, he was put in charge of the exploration of Bengal, a vast territory twice the size of France. He left in 1807 and, until 1814, carried out numerous expeditions on foot, and occassionally on the backs of elephants. His health waning, he decided to return to his native Scotland with his collections and illustrations, but he was informed that he could not take them as they belonged to the Company. Buchanan felt obliged to leave them to the care of Nathaniel Wallich (1786–1854), director of the botanical garden of Calcutta.

Buchanan returned to Scotland in 1816. Two years later, the death of his brother resulted in the inheritance of significant property belonging to his mother. He decided to join the name associated with these, Hamilton, to his own name. From then on he devoted himself to the study of the Indian flora and fauna, referring to the collections kept by the Linnaean Society of London. He described a large number of species of plants, fish and mammals.

The illustrations left at Calcutta would only be rediscovered by Edward Blyth (1810–1873) after Buchanan-Hamilton's death. They featured 378 birds, some species for the first time. Blyth later dedicated the Grey-necked Bunting (*Emberiza buchanani*) and the Rufous-fronted Prinia (*Prinia buchanani*) to Buchanan-Hamilton.

Thomas Claverhill Jerdon (1811–1872) studied medicine in the renowned faculty at the University of Edinburgh, and became a surgeon in the East India Company in 1835. He ended his career as Chief Hospital Inspector. He very quickly started to amass a collection of birds, which he forwarded to Scottish ornithologist Sir William Jardine (1800–1874) to study. Unfortunately, his parcel was attacked by insects in transit from India to Great Britain. What's more, Jardine, concerned with contaminating his own collection, refused to receive Jerdon's parcels. The latter then decided to study the birds that he collected himself. He released several publications on Indian avifauna in which he mentioned 420 species, double the list established by Colonel Sykes. After having published the four parts of *Illustrations of Indian Ornithology* from 1843 to 1847, he published a major work, *The Birds of India*, from 1862 to 1864. This work, in which more than 1,000 species were

Thomas Claverhill Jerdon (1811–1872), physician in the service of the East India Company and author of important publications on Indian avifauna.

described, would be reprinted 12 times. Jerdon was not just a designator of species, he took it upon himself, as much as possible, to describe the habits of the birds and mammals that he studied.

- The scientific institutions of the colonies

The East India Company founded a botanical garden in Calcutta in 1776 with the goal of studying and improving the growth and acclimatization of useful plants. Around 1800, Lord Wellesley created a menagerie so that the vertebrates of the Indian subcontinent described by naturalists could be subjected to more detailed observations. Towards the middle of the century, the first local scholarly societies began to emerge. It was in one of these societies that **Edward Blyth** (1810–1873) would distinguish himself. Having lost his father at a young age, he studied at Wimbledon School. A mediocre student, he preferred to explore the woods and fields of the surrounding area. At 15 years of age, he began to study chemistry with the objective of attending university and joining the church, but his passion for natural history often motivated him to rise at three or four in the morning to observe and gather insects or birds. As an adult, he used his modest

Edward Blyth (1810–1873) was one of the founders of Indian zoology. He maintained a long correspondence with Charles Darwin (1809–1882).

The Bar-tailed Tree-creeper (*Certhia himalayana*), by Khuleelooddeen, an Indian artist, painted for an account by Edward Blyth.

inheritance to buy a small hardware shop in the suburbs of London. He showed little interest in his business and took a room in the centre of London to be closer to the British Museum and the libraries. He started to publish his ornithological observations on a regular basis. His first publications won him praise for the quality of his observations, but also some criticism for his lack of reserve in changing the names of birds he thought were not suited to them. It was following the criticism expressed by Hugh Edwin Strickland (1811–1853) that Blyth recognized the importance of rigour in naming and started to adhere to the code of zoological nomenclature.

Unsurprisingly, Blyth's business soon failed. He became a curator for the Zoological Society of London for some time, but his health was failing and he was advised by his physician to move to a warmer climate. He was then offered the position of curator of the museum of the Asian Society of Bengal. He arrived in Calcutta in September 1841. Blyth devoted the next two decades to the fauna of the region and greatly supplemented the collections of the institution. In 1840, he published the catalogue of the museum's birds and launched the society journal.

Despite poor health and a meagre income (his salary remained the same throughout his stay in India), he was able to conduct some interesting expeditions. He also received specimens from several specialists in the region, including Hume and Tickell, as well as specimens from Taiwan and China sent by Swinhoe.

At 44 years of age he married a widow he met in England. Her death after only three years of marriage, as well as his mounting financial problems, convinced him to leave the country. He continued to publish scientific and popular articles. In recognition of the quality of his work, the British Ornithologists' Union named him an honorary member and later a distinguished member.

- The members of the colonial administration

The colonization of India by the United Kingdom gave many of the members of the administration the opportunity study the fauna of the subcontinent.

Brian Houghton Hodgson (1800–1894) was one of the first ethnologists of Tibet. His work on birds helped complete the work of the ornithological pioneers. Hodgson's father had lost all of his fortune to risky investments. At 17 years of age, Brian Hodgson entered into the service of the English East India Company and immediately left for Calcutta to study the

Although ornithology was nothing more than a hobby for Brian Houghton Hodgson (1800-1894), he exerted a lasting influence on the discipline. His name remains primarily associated with the introduction of Buddhism to Europe.

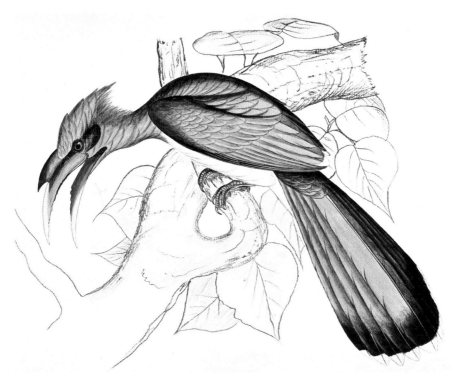

An Indian Grey Hornbill (*Ocyceros birostris*), painted by an Indian artist for an account by Brian Houghton Hodgson (1800–1894) between 1820 and 1858.

local languages and law. However, his health was poor and he suffered from fever and an affliction of the liver. The Company's physician gave him the choice between likely death if he remained and returning to England (and thus being declared unsuitable for service). The young Hodgson deemed death a less terrible fate than returning to his country as an invalid and being in the care of his family. He would fight his sickness for the rest of his life in order to be able to continue his duties.

He demonstrated a brilliant aptitude for the study of foreign languages, particularly Persian and Sanskrit. He was then sent to Kumaon, a mountainous region north of Delhi that had just been conquered by the British. His responsibilities here included taking inventories of each community in the region.

In 1822, he received a promotion that required him to relocate to Calcutta. However, he could not bear the climate and his health quickly deteriorated. He then left for Kathmandu, where he remained for the next 20 years. Though he had had no formal scientific training, he developed an interest in natural history. He assembled specimens and had many illustrations rendered by talented local artists. In this way, he built up a collection of 1,241 paintings of birds, often accompanied by illustrations of their nests and eggs, as well as notes describing their measurements, breeding behaviour and

Allan Octavian Hume (1829-1912) actively fought for the Indian people to play a role in the government of their country, the first step towards independence. He is considered the father of Indian ornithology.

Title page of *The Nests and Eggs of Indian Birds* (1883), one of the works that Allan Octavian Hume devoted to the birds of India.

their diet (determined by examining the contents of their stomachs). He forwarded it all to Allan Hume and his collections would be bequeathed to the Zoological Society of London.

His work as a naturalist did not end there. In addition to the new species of reptiles and mammals that he discovered (including the Tibetan Antelope, *Pantholops hodgsonii*), he greatly enriched knowledge of the avifauna of the region. His collection of more than 9,500 bird skins was sent to the British Museum. It consisted of 972 species of which 124 were new. Fluent in Nepali, he knew the area and its residents well. A friend of the Dalai Lama, he would publish several books on the region and contributed to making Buddhism known in Europe.

Allan Octavian Hume (1829-1912), son of a Scottish politician, studied medicine before leaving for India in 1849. Employed by the colonial administration in Bengal, he created a journal in Bengali. He participated in the repression of the Indian revolt in 1857 and quickly climbed the ranks of the colonial administration. He was interested in education and was one of the figures behind the push to provide free education in India and the creation of an educational system for juvenile delinquents. He also fought against alcoholism and infanticide and was an activist for the education of women. He did not hesitate to criticize the choices of the British government, whom he accused of wanting to make India into a second Great Britain without any concern for the aspirations and well being of Indians. He left his duties in 1882 and the following year would incite the graduates of the University of Calcutta to form their own political movement. In 1885, he became secretary general of Indian National Congress.

Passionate about science, Hume began to amass a rich collection of birds from the Indian subcontinent. He would publish numerous works on this fauna. Despite the destruction of a portion of his collections after a landslide, he left a total of 80,000 specimens to the British Museum, including 258 biological types.

He described a large number of species, including Hume's Owl (*Strix butleri*), Plain Leaf Warbler (*Phylloscopus neglectus*) and Mongolian Ground-Jay (*Podoces hendersoni*). In 1872, he began to publish *Stray Feathers, A Journal of Ornithology for India and its Dependencies* to which he was one of the primary contributors. He was also the author of *The Nests and Eggs of Indian Birds* (1883) and *Game Birds of India, Burma and Ceylon* (1879).

- The exploration of China

It was a member of the staff of the British Consulate who would receive the first descriptions of the avifauna of China's coastal regions. **Robert Swinhoe** (1836–1877) shared the title of pioneering naturalist of China with French missionary Père Armand David (1826–1900), and the Russian general Nikola Mikhalovitch Prjevalski (1839–1888). Swinhoe became an interpreter in Hong Kong at 18 years of age. The following year he was transferred to the consulate in Amoy (modern day Xiamen), on the Chinese coast. He published his first work on the fauna of the region three years later. He served as the interpreter of Sir James Hope Grant (1808–1875) and General Robert Napier (1810–1890) during the second Opium War in China. The military expedition came to an end with the fall of Beijing. Swinhoe then took an interest in the markets of the city: the hunters were so numerous that simply studying the contents of the stalls allowed the naturalist to form a fairly accurate picture of the surrounding avifauna. After the war, in 1861, he became a vice-consul in Taiwan. There, he could easily study the fauna and flora of the island and sent many specimens to the British Museum. He constructed an aviary, which allowed him to keep the rare Swinhoe's Pheasant (*Lophura swinhoii*) and Whistling Green-Pigeon (*Treron formosae*). The first specimen of this bird was brought to him by a Chinese hunter, plucked of its feathers. Shortly thereafter a collector employed by Swinhoe brought him a specimen with all its plumage.

His return to Amoy would allow him to resume his explorations and he soon discovered a black marine bird, *Oceanodroma monorhis*, which was later named Swinhoe's Storm-petrel. In 1868, he participated in a pirate hunt aboard a gunboat, during which he assembled a collection of 172 species of birds, 19 of them new to science. He was later overtaken with paralysis, which prevented him from continuing his expeditions. Living in Shanghai during the winter season, he made his way to the markets every day to discover rare or novel species. His health would get so bad that in 1875 he was forced to return to London. He continued to work on his collections, often with the help of other naturalists. Curiously, these collections, containing 3,700 skins representing 650 species of 200 types, were refused by the British Museum and eventually the bulk of them were acquired by Henry Seebohm (1832–1895). He would in turn bequeath his personal collections, including Swinhoe's, to the British Museum.

A large part of the collection and research undertaken on birds in the overseas territories was financed by governments. The states directly financed research institutions and universities

Robert Swinhoe (1836–1877) was the first Westerner to explore Taiwan. As an early settler, like Père David in China and Edward Blyth in India, he discovered numerous new species.

and promoted the organization of expeditions. Governments also played an essential but indirect role by maintaining military and administrative personnel. This proved to be a huge advantage over expeditions because staff remained in the same area for extended periods (sometimes an entire lifetime), which allowed for more extensive and exhaustive collections than those gathered during expeditions.

- **The ecclesiastics**

This overview would not be complete without including the men of the church, who were very well represented amongst the first naturalists. Of the 760 authors cited in *Bibliography of British Ornithology* by Harry Kirke Swann (1871–1926), 60 are ecclesiastics, a phenomenon that was common to all branches of natural history. In Europe, Abbot Octave Aubry, vicar at Saint-Louis-en-l'Isle in the Dordogne, assembled the most extraordinary collections of the early 19th century (with specimens from the Indian Ocean, India and the Philippines), the Jesuit Francesco Cetti (see page 89) studied the avifauna of Sardinia, pastors C.L. Brehm (see page 120) and Otto Kleinschmidt (1870–1954) both built up impressive collections of German birds, Reverend Tristram (see page 154) played a prominent role in British ornithology, and Reverend Francis Jourdain (1865–1940) assembled the largest collection of bird eggs in Europe.

The work of missionaries would also be essential to the study of birds. Their numbers, significantly reduced in the early 19th century, would grow in the decades that followed, in step with European colonization. They were often the first to enter into unexplored regions. The observations that they made, like the specimens they gathered, were of the utmost importance. One of the best examples is that of **Père Armand David** (1826–1900), a French Lazarist missionary who took part in many voyages to China between 1866 and 1872. He was the first to describe the Giant Panda (*Ailuropoda melanoleuca*), the Buddleia or Butterfly-bush (*Buddleja davidii*) and Père David's Deer (*Elaphurus davidianus*). With Émile Oustalet (1844–1905), he published *Les Oiseaux de la Chine* (1877) in which he stated that he had seen 772 different species of birds. He sent more than 1,300 skins of 470 species to the Muséum de Paris.

The role of missionaries would slowly decline around the turn of the century, with increasing competition from organized expeditions conducted by scientific institutions (particularly American ones), and from professional hunters and collectors.

The name of Père Armand David (1826–1900) is associated with the discovery of the Giant Panda.

Officers, physicians, ecclesiastics, and travelling naturalists; all of these men contributed to the formation of networks that facilitated the exchange of specimens and efficiently relayed information to the West.

The progressive organization of ornithology in Europe

Natural history was one of the great passions of this century. Nearly everyone collected insects, shells or animal skins in some form. Attendance at public conferences and subscriptions to courses on natural history very often exceeded capacity, and botanical outings were popular. It was a mark of good taste to be an amateur naturalist. Two examples demonstrate this popularity: the proliferation of works on regional fauna and the creation of specialized learned societies. This phenomenon was not restricted to ornithology and took place in many branches of natural history. Gilbert White's wish (see page 85) that "each kingdom, each province, owes it to itself to possess its own monograph" would very quickly be fulfilled.

• The founding work of the Naumanns and Brehm

The lack of large reference collections in Germany was compensated for by the creation of several museums at the beginning of the century (see page 103). Outside of these institutions, the work of several naturalists of modest means contributed to the blossoming of German ornithology. **Johann Andreas Naumann** (1744–1826) was a farmer with land 160 kilometres south-west of Berlin. This marshy and humid region was not very well suited to agriculture, but attracted a richly varied avifauna. A passionate hunter, Naumann was also a keen observer and began to keep a meticulous record of his observations. He published several small books on the birds of the region. Two of his sons would inherit his interest for birds and the hunt. The eldest, **Johann Friedrich Naumann** (1780–1857), interrupted his studies at 14 years of age to work on the family farm. A brilliant illustrator, he began to illustrate his father's works. The father and son then published *Naturgeschichte der Vögel Deutschlands* from 1820 to 1844. If Naumann's scientific knowledge was somewhat lacking, his simple, accessible style and his excellent understanding of the habits of birds, would lead to widespread popular success. These publications would inspire numerous young Germans to take an interest in ornithology.

Johann Andreas Naumann (1744-1826) (above) published several books on the birds he observed around his farm in Prussia. His son, Johann Friedrich Naumann (1780-1857) (below) continued his father's work. A museum commemorating this family of ornithologists was created in Köthen castle.

A chromolithograph of the Great Auk (*Pinguinus impennis*), by Wilhelm Blasius, based on a painting by J.F. Naumann, from *Naturgeschichte der Vogel Mitteleuropas* (1895–1905). The image is an artist's impression, as the species became extinct in the middle of the 19th century.

The publications of Christian Ludwig Brehm (1787–1864) encouraged many amateurs to take up the study of birds.

J.F. Naumann and C.A. Buhle (1773–1855) would publish a work on eggs, *Die Eier der Vögel Deutschlands*, from 1818 to 1828. They called on Christian Ludwig Nitzsch (1782–1837) to fill in the gaps in their anatomical knowledge. The success of J.F. Naumann's works led to a number of honours, including the title of professor of natural history, conferred to him by the Duke of Köthen.

A famous pedagogue who corresponded regularly with numerous amateurs, **Christian Ludwig Brehm** (1787–1864) would influence an entire generation of ornithologists. Following his studies at the University of Jena, he became a priest at Renthendorf where he spent the remainder of his life. Like the Naumanns, he was a keen observer of birds in their environment. He was the author of numerous publications, one of the most important of which was without a doubt *Beiträge zur Vögelkunde in vollständigen Beschreibungen mehrer neu entdeckter und vieler seltener oder nicht gehörig beobachteter deutscher Vögel* (1820–1822), in which he meticulously described 104 species of birds. However, his attempt to launch a periodical devoted to birds, *Ornis*, turned sour. Throughout his life, Brehm would try in vain to sell his collection to the museum in Berlin. Finally, after being stored in an attic for nearly 40 years, it would be sold by Otto Kleinschmidt (1870–1954) to Lord Rothschild (see page 156). Brehm's son was the renowned zoologist Alfred Edmund Brehm (1829–1884), author of numerous well-known popular works.

During this century, Germany would be home to a great number of significant ornithologists, such as Johann Matthäus Bechstein (1757–1822), author of an important manual of German birds, Gustav Hartlaub (1814–1900), author of works on the birds of Africa, and Heinrich Gätke (1814–1897), a specialist in bird migrations.

In 1845, roughly 20 German ornithologists gathered under the presidency of J.F. Naumann and decided to publish a journal dedicated to ornithology, *Rhea*, whose direction was entrusted to Ludwig Thienemann (1793–1858). The meetings that followed led to the creation of an ornithological society, the Deutsche Ornithologen-Gesellschaft (DO-G) in 1851. Directed by J.F. Naumann, it comprised 107 members in 1854 and 230 in 1858.

Rhea, which would be scrapped after only two editions, was replaced by *Naumannia*, edited by Eduard Baldamus (1812–1893). Very quickly differences of opinion were voiced by Jean Louis Cabanis (1816–1906), one of the foremost German ornithologists. In 1853 Cabanis would launch *Journal für Ornithologie,* which would be chosen as the official communication vehicle of the DO-G. However, the quarrels continued and Cabanis decided to create a new society in 1867, the Deutsche Ornithologische Gesellschaft (DOG). Eventually the two organizations, the DO-G and the DOG, would merge into a single body, the *Allgemeine Deutsche Ornithologische Gessellschaft*, inaugurated in 1875. Cabanis's publication would be bought back in 1894 and the society remained active until the Second World War.

History would repeat itself during the Cold War when the DOG was active in West Berlin and a DO-G was launched in East Germany. Legal issues would not permit their merger after the fall of the Berlin Wall and the two societies continue to co-exist today.

- British ornithology

At the very beginning of the 19th century, the *Ornithological Dictionary or Alphabetical Synopsis of British Birds* (1802) and its *Supplement* (1813), by **George Montagu** (1753–1815) would not attain the popularity of the work of his predecessor

Title pages of *Journal für Ornithologie* (above) by Jean Cabanis and *Naumannia* (below) by Eduard Baldamus.

Miniature portrait on ivory of George Montagu (1753–1815).

SUPPLEMENT

TO THE

Ornithological Dictionary,

OR

SYNOPSIS OF BRITISH BIRDS.

BY

GEORGE MONTAGU, Esq. F.L.S. & W.W.S.

Printed by S. WOOLMER, EXETER;

AND SOLD BY E. BALISTER, 88, STRAND, T. AND J. ASTIL, CORNHILL, AND THOMAS UNDERWOOD, 32, FLEET-STREET, LONDON; BY WHOM MAY BE HAD "TESTACEA BRITANNICA, OR BRITISH SHELLS" INCLUDING THE FRESH-WATER, LAND, AND MARINE, WITH DESCRIPTIONS, AND REFERENCES TO PLANTS, AND FIGURES; ALSO, "ORNITHOLOGICAL DICTIONARY," IN TWO VOLS, BY THE SAME AUTHOR.

1813.

Title page of *Supplement to the Ornithological Dictionary* (1813) by George Montagu. His works ushered in a new standard for quality: the descriptions are more precise and facts verified. Not confined to a single field of study, he also published works on molluscs and gunpowder.

Logo of the British Ornithologists' Union.

Thomas Bewick (see page 87), but brought a higher standard of precision and rigour. Montagu was interested in the variations of birds, whether they were linked to the seasons, sexes or maturity. He did not hesitate to breed species in order to gain a better understanding of changes in their appearance. It was commonly believed that the winter and spring forms of the Ruff (*Philomachus pugnax*) and of the Red Knot (*Calidris canutus*) were in fact two different species. His breeding demonstrated that they were in fact a single species in both cases. His prudence led him to omit the Black Woodpecker (*Dryocopus martius*), which he had never himself observed in England, from his book. He also noted that the last English populations of the Great Bustard (*Otis tarda*), would soon disappear owing to the fact that the young, which were incapable of flight, were the easy prey of sheepdogs. When he published his supplement 11 years later, Montagu indicated that shepherds had not observed a single specimen of this bird for several years. In 1831, James Rennie (1787-1867) published an improved version of Montagu's dictionary; Edward Newman (1801-1876) would make further additions in a version published in 1866.

The rise of a new generation of ornithologists in Great Britain led to the publication of two major works under the same title in 1837: *History of British Birds*. The first was the work of William MacGillivray (1796-1852) and the second of William Yarrell (1784-1856). The first of these was arguably the most original and precise work, although MacGillivray's classification was somewhat controversial. The second would be the standard reference on British avifauna for several decades. Yarrell was a correspondent of Bewick and friend of John Gould, William Jardine and numerous other naturalists. The works would be reprinted countless times in 1845, 1856 and 1871, and revised by Howard Saunders (1835-1907) in 1885. Yarrell's collections and library were divided up when they were sold at auction.

British ornithologists would very quickly follow the example of their German colleagues. The **British Ornithologists' Union** (BOU) was founded in November 1858 by 12 people, including Reverend Tristram, Frederick DuCane Godman (1834-1919), Sir Edward Newton (1832-1897), Alfred Newton (1829-1907), P.L. Sclater, and Wilfrid Hudleston Hudleston (1828-1909). Its first president was Colonel Henry Maurice Drummond-Hay (1814-1896).

The journal of the society, *The Ibis*, was founded in 1859 by **Philip Lutley Sclater** (1829-1913). In addition to his contribution to the success of the BOU, Sclater was the author of important works on zoogeography. In 1858, he defined six zoological regions: Palaearctic, Ethiopian, Indian, Australasian, Nearctic and Neotropical. These biogeographic zones are still in use today.

In 1889, another organization was founded: the Royal Society for the Protection of Birds (RSPB). Its goal was to fight the trade of feathers and leather of the Great Crested Grebe (*Podiceps cristatus*), used in the women's wear industry. Its populations were so overhunted that it was estimated that there were only 1,000 pairs remaining in Great Britain and Ireland when the RSPB was formed. This society was not the first to take on the protection of animals. That honour goes to the Royal Society for the Prevention of Cruelty to Animals founded in 1824. However, it was the first to fight against the destruction of the natural environment as well as cruelty towards animals.

- French ornithology

French ornithology in the 19th century would not share the dynamism of its neighbours. We can however note the publication of numerous works, including: *Tableau élémentaire d'ornithologie, ou Histoire naturelle des oiseaux que l'on rencontre communément en France* (1806, reprinted in 1822) by Sébastien Gérardin (1751–1816); *Ornithologie française ou Histoire naturelle générale et particulière des oiseaux de France* (1823), by L.J.P. Vieillot; *Ornithologie du Gard et des pays circonvoisins* (1840) by Jean Crespon (1797–1857); *Ornithologie de la Savoie, ou Histoire des oiseaux: qui vivent en Savoie à l'état sauvage soit constamment, soit passagèrement* (1853–1854) by Jean-Baptiste Bailly (1822–1880); *Essais étymologiques sur l'ornithologie de Maine et Loire: ou les moeurs des oiseaux expliquées par leurs noms* (1859) by Michel Honoré Vincelot (1815–1877); *Ornithologie parisienne: ou catalogue des oiseaux sédentaires et de passage: qui vivent à l'état sauvage dans l'enceinte de la ville de Paris* (1874) by Nérée Quépat (1845–1927) and *Ornithologie de la Sarthe* (1880) by Ambroise Gentil (1842–1929).

The work of painter and naturalist **Jean Louis Florent Polydore Roux** (1792–1833), who published *Ornithologie provençale* from 1825 is also noteworthy. He exhibited paintings at the Salon de Paris on several occasions and became the curator of the Muséum de Marseille in 1819. He was interested in various subjects and published *Catalogue d'insectes de Provence* (1820) and *Histoire naturelle des crustacés de la Méditerranée* (1828). Roux decided to accompany Baron von Hügel (1796–1870) who was leaving for India in 1831, and would perish two years later when he came down with what appeared to be the plague while exploring the Himalayas. His work on shellfish

Title page from *Ornithologie provençale* by Polydore Roux (1825).

ORNITHOLOGIE
PROVENÇALE,

Male Wallcreeper (*Tichodroma muraria*), from *Ornithologie provençale* by Polydore Roux.

and his *Ornithologie* would remain unfinished. Only about 450 plates would be published in two volumes, as well as the text of the former.

French ornithology, which included strong personalities like Levaillant and Vieillot, would organize itself much later than its German, British and American counterparts. The first French ornithological journal would only be published in the early 20th century thanks to the efforts of Louis Denise (see page 203). It is very difficult to determine the reasons for this situation, as France had no fewer passionate amateur ornithologists than its neighbours. It is likely that the extreme centralizing power of the Muséum National d'Histoire Naturelle played a role (until 1900, it received the largest portion of the research budget given out by the French government, ahead of the Collège de France, École Normale Supérieure and the science faculties in Paris). Moreover, it must be noted that the provinces played host to a great number of other, often generalist, learned societies, including Linnaean Societies (in Lyon, Bordeaux, and Normandy). Bird breeders gathered under the banner of the Société Française d'Acclimatation. Nevertheless, these factors did not prevent the formation of other specialized societies, like the Société Entomologique de France in 1832 or the Société Botanique de France in 1854.

The rise and influence of American ornithology

America would become the world's most flourishing centre for ornithology in the 19th century (today the most important ornithological collections are almost exclusively housed in the United States, see page 138). Beside the observations gathered in the first explorations, the serious scientific study of America began with Mark Catesby (see page 63), William Bartram (see page 95) and John Abbot (see page 127).

While European ornithology was concentrated in a few large institutions (London and Tring in the United Kingdom, Paris in France, Leiden in the Netherlands, Vienna in Austria, Berlin in Germany), American ornithology was supported by the sudden appearance of numerous institutions in different parts of the country, including the Academies of Science of Philadelphia and California, the museums of Washington, New York, Chicago and Harvard, and by the specialized instruction in a number of universities that conducted their

The American Passenger Pigeon (*Ectopistes migratorius*) suffered a rapid decline due to humans. Its population was estimated at more than two billion in 1810. The last known individual died at the Cincinnati Zoo in 1914. Like the bison, this bird was symbolic of the relentless overexploitation of natural resources.

Great Heron

own research. The American government backed the study and protection of fauna, and notably game, through the establishment of a biological research service in 1886. Finally, three great ornithological organizations appeared over the course of the 19th century, which would bring together amateurs, coordinate scientific study and promote the publication of works on birds.

The rapid modernization of America was accompanied by a strong sense of nostalgia: many Americans were under the impression that the wilderness was disappearing and with it, a certain ideal of beauty and liberty. The need for a government policy to conserve natural sites and their inhabitants was developed in 1832 by George Catlin (1796–1872), famous for his paintings of Native Americans. Yellowstone Natural Park was inaugurated in 1872. The urban populations would take a keen interest in the study of fauna and flora. Books on birds became bestsellers: *Birds and Poets* (1877) by John Burroughs

(1837–1921) sold more than a million copies, *Bird Guide* by Chester Albert Reed (1876–1912) more than 600,000 copies, Ernest Thompson Seton's (1860–1946) books on birds and mammals sold nearly two and a half million copies. The protection of nature and the study of behaviour would allow many women to play a significant role, another unique feature of American ornithology.

- The forerunners: Abbot and Wilson

London-born **John Abbot** (1751–1840 or 1841) took an interest in insects at a very early age. Encouraged by his father to collect and draw them, he quickly proved to be a very talented illustrator: some historians consider the drawings he completed during his adolescence among the best natural history illustrations of the period. From 18 years of age, he worked as a clerk in his father's office. In 1773, having no desire to remain in this line of work, he sold his collection of insects and left for the American colonies. A private group of naturalists headed by Dru Drury (1725–1804), put him in charge of gathering the natural history specimens of the New World. He first settled in Virginia; however, disappointed with what he could find, he moved to Georgia where he would stay for the rest of his days. He did not publish any works on his own, but sold his specimens and watercolours to European naturalists. It is thus that his descriptions and specimens would turn up in many ornithological works by John Latham (see page 82).

It was with the contribution of **Alexander Wilson** (1766–1813) that American ornithology really took flight. The son of humble Scottish farmers, his parents' wishes would have seen him join the priesthood. An apprentice weaver at the age of 13, he would himself become a weaver and travelling merchant in 1782. He was drawn to poetry and published a collection of poems in 1790, and another in 1791. His texts caused him some troubles, as they described the difficult life of weavers and the unjust treatment that they were subjected to by their employers. Forced to burn his books and pay fines, he chose to leave Scotland and made his way to the United States in 1794. He became a schoolteacher and, in 1802, met naturalist William Bartram (1739–1823), who gave him access to his library. He then conceived a project to publish an illustrated work on the birds of North America. He began to travel the country to carry out observations and gather subscriptions for his upcoming publication. The first volume of *American Ornithology* was published in 1808, the sixth and final one, in 1812. Two more volumes would be published after his death in

John Abbot (1751–1840 or 1841) is primarily known for his natural history illustrations.

Alexander Wilson (1766–1813) is, together with Mark Catesby and John James Audubon, considered to be one of the three founders of American ornithology.

Above, an American Goldfinch (*Carduelis tristis*), centre, a Blue Jay (*Cyanocitta cristata*), below, a Baltimore Oriole (*Icterus galbula*), from *American Ornithology* by Alexander Wilson.

Title page of *American Ornithology* (1828 edition) by Alexander Wilson.

1814, thanks to the efforts of his friend George Ord (1781–1856). Wilson expressed horror at the widespread destruction of passerines (often for food), notably the American Robin (*Turdus migratorius*), Bobolink (*Dolichonyx oryzivorus*) and Red-winged Blackbird (*Agelaius phoeniceus*), but his indignation would long remain unheard.

The name of **Thomas Nuttall** (1786–1859) is associated with the Nuttall Ornithological Club (see page 137). Founded in 1873, it was the first scholarly society dedicated to the birds of the Unites States. Born in Yorkshire, Nuttall became an apprentice in his uncle's printing works in Liverpool, then in London. Already having developed a keen interest in natural history, he was supported by Philadelphia-based botanist Benjamin Smith Barton (1766–1815) from the moment he arrived in America in 1808. The latter encouraged him to travel to study the fauna and flora of the country. Thomas Nuttall made his first trip to Delaware in 1809, followed by many more until 1842, when he returned to Great Britain.

Nuttall published a large number of works on the plants and geology of the country, as well as a *Manual of the Ornithology of the United States and Canada*. The first volume dealt with terrestrial birds (1832, revised edition in 1840) and the second with aquatic birds (1834, reprinted and updated several times until 1903). This small, inexpensive guide contributed to the popularization of ornithology in the United States. He described or collected many new species, including the American Black Oystercatcher (*Haematopus bachmani*), Western Gull (*Larus occidentalis*), Townsend's Solitaire (*Myadestes townsendi*), Green-tailed Towhee (*Pipilo chlorurus*), Harris's Sparrow (*Zonotrichia querula*), Anna's Hummingbird (*Calypte anna*), and Tricoloured Blackbird (*Agelaius tricolor*).

Thomas Nuttall (1786–1859) was a remarkable naturalist who undertook a number of expeditions throughout North America.

A member of several important learned societies (Linnaean Society of London, the Academy of Natural Sciences of Philadelphia and others), Nuttall played a pioneering role in a number of fields: botany, zoology, ecology and horticulture. His many explorations largely surpassed those undertaken before him and the quality of his observations would win him much praise: for Audubon "no one knew how to describe the songs of our different species of birds like Nuttall".

● The prince of ornithology: Charles Lucien Bonaparte

Charles Lucien Bonaparte (1803–1857) was the nephew of Emperor Napoleon. His parents, Lucien Bonaparte and Alexandrine de Bleschamp, widow of a Parisian banker, married in secret. Napoleon, who would have preferred to see his brother marry nobility, tried to persuade him to divorce. The family took refuge in Rome under Pope Pius VII. When the Emperor kidnapped and imprisoned the Pope in 1810, the family tried to emigrate to North America, but the ship was required to berth in Sardinia where they were taken prisoners by the British. They were later kept under surveillance in England, near Grimley. There, the Bonaparte family formed ties with that of naturalist Thomas Andrew Knight (1759–1838). The young Charles discovered natural history through his collections. When peace was declared in 1814, the family left for Italy, where the Pope bestowed the title of Prince of Musignano on the young Charles.

Charles Lucien Bonaparte (1803-1857), nephew of the French Emperor, spent many years in the United States.

Bonaparte devoted his time to the study of science and building a natural history collection. He referred to the first two volumes of Coenraad Jacob Temminck's (see page 142) *Manuel d'ornithologie* to identify birds, and sent him a specimen

Young male Andean Condor (*Vultur gryphus*), from *American Ornithology* by Charles Lucien Bonaparte.

Title page of the first volume of *American Ornithology* (1825) by Charles Lucien Bonaparte.

of a species he had hunted several years earlier and had never been able to identify. Temminck rightly noted that it was a new species, the Moustached Warbler (*Acrocephalus melanopogon*).

Bonaparte would marry his cousin Zénaïde in 1822 and received an endowment of 730,000 francs from his father-in-law, Joseph Bonaparte, the former King of Naples and Spain. The young couple left to settle near Philadelphia, where Zénaïde's father owned property.

He did not waste time in getting involved in the active scientific life of Philadelphia and quickly became a member of the Academy of Natural Sciences of Philadelphia. He befriended the artist Titian Ramsay Peale (1799–1885), entomologist Thomas Say (1787–1834) and naturalist William Cooper (1798–1864). Bonaparte then took on the revision of Alexander Wilson's (1766–1813) work, *The American Ornithology* under the title *American Ornithology, or the Natural History of Birds Inhabiting the United States, not Given by Wilson*, which commenced publication in 1825 (the second and third volumes were published in 1828, the last in 1835). The works of Wilson and Bonaparte would be published in many combined editions. Bonaparte's observations were conducted almost exclusively from the study of preserved specimens and he provided very little in terms of field observations.

In 1828, he published 'The Genera of North American Birds, and a Synopsis of the Species found within the territory of the United States; systematically arranged in Orders and Families', an article in which he introduced an important innovation in classification: he used the presence or absence of a hidden toe to replace the more typical division between aquatic and terrestrial birds.

Bonaparte's contribution marked an essential step in the development of American ornithology because he clarified the taxonomy of genera and species, his familiariarity with the fauna of Europe proving very helpful in this regard. In the mid-1820s, he began to conceive a very ambitious project: the study of the bird genera of the entire world. In 1826, dwindling finances forced him to leave for Italy in the hopes of recovering a portion of his wife's dowry. American naturalists deeply lamented his departure. He would only return to America one more time, in 1827, to prepare the second and third volumes of his *American Ornithology*.

He settled in Rome in 1828, remaining there for nearly 20 years, and dedicating himself to the study of the avifauna of the Italian peninsula. He would publish three volumes of *Iconographia della Fauna Italica* between 1832 and 1841. Little by little he became more involved in politics, and took on the vice presidency of the legislative counsel of the Roman constituent assembly in 1848, signing a proclamation calling on Romans to take up arms. The French army took siege of the city, which would soon fall. A new exodus ensued and drove him to Marseille, then Paris. He ignored Louis Napoleon's order to immediately leave the French territory. Stopping at Orléans, he then went to Le Havre and on to England, where he took part in the scientific life of London and made contact with Sir William Jardine (1800–1874). He stayed in Leiden for a year, where his friend Hermann Schlegel (see page 144) allowed him to study his rich bird collection of 12,000 specimens at his leisure. The two men travelled to Germany, where they examined the collection of Hinrich Lichtenstein (1780–1857), then met up with Johann Friedrich Naumann (see page 119). In the summer of 1850 Bonaparte received a letter from Louis Napoleon authorizing his return to France. He then settled in Paris, where he spent the rest of his days. He considered his cousin to be a traitor of the Republic, and he would avoid the imperial court for the rest of his life.

1. Steller's Jay.
Garrulus Stelleri.

2. Lapland Longspur. 3. Female.
Emberiza Lapponica.

Above, Steller's Jay (*Cyanocitta stelleri*), below, male and female Lapland Bunting (*Calcarius lapponicus*), from *American Ornithology* by Charles Lucien Bonaparte.

In 1850, he published *Conspectus generum avium*, followed by a second volume in 1857, the year of his death. Despite a rather tormented life, he left an indelible mark on the ornithology of the first half of the 19th century by virtue of his great understanding of the fauna of the New and Old World.

- The first American-born ornithologists

American science would not break its dependence on Europe until the middle of the century. Up until then the naturalists who were active in the United States, such as George Edwards, Thomas Pennant, John James Audubon, William Swainson, Paul du Chaillu and many others, came from the Old World. Little by little, they gave way to ornithologists born on American soil.

John Cassin (1813–1869) was a good example of these new scientists. The son of farmer, born in Pennsylvania, Cassin began life as a shopkeeper before taking over the leading lithography and engraving business in Philadelphia. Passionate about natural history from a young age, he joined the city's Academy of Natural Sciences in 1842 where he assumed the position of curator until 1869. He began to study the Academy's extensive collection of birds, strengthened by the 26,000 specimens bequeathed by Thomas Bellerby Wilson (1807–1865). His publications would shape Cassin into the leader of American ornithology: his most celebrated works were *Illustrations of the Birds of California, Texas, Oregon, British* and

In addition to the study of Wilson's collection, John Cassin (1813–1869) (below) was responsible for the study of the birds brought back from several famous American expeditions, including those of Charles Wilkes (1798–1877) in the Pacific and Arctic.

Collared Kingfisher (*Todiramphus chloris*), plate by John Cassin.

The Common Raven (*Corvus corax sinuatus*), from *The Birds of North America; the Descriptions of Species Based Chiefly on the Collections in the Museum of the Smithsonian Institution* (1860), by Spencer Fullerton Baird.

Bowen & Cº Lith & Col Philada

Spencer Fullerton Baird (1823-1887) was one of the great patrons of 19th century science.

Russian America (1853-1856) and *Birds of North America* (1860). Cassin was presented the project of creating a museum for the Academy, an institution that would rival those of Europe. His approach consisted of buying large collections, like that of French collector the Duke of Rivoli (15,000 specimens), the Australian birds of John Gould (2,000 specimens) and Marc Athanase Parfait Oeillet Des Murs's (1804-1878) collection of eggs. Thanks to his efforts, the museum possessed the world's largest ornithological collection, with 29,000 specimens. In 1861, the American Civil War broke out and Cassin enlisted in the armies of the North. His detention in the confederate prisons would have a profound affect on his health, and put an end to his scientific career.

Spencer Fullerton Baird (1823-1887) would play a significant role in the organization of scientific research in the United States and would influence the careers of many ornithologists. At 17 years of age, he entered into correspondence

The exploration of North America

The expedition conducted by Meriwether Lewis (1774-1809) and William Clark (1770-1838) was the first great scientific expedition financed by the young American nation. It traversed the United States and reached the Pacific after three years (1804-1806). The chief objective of the expedition was to identify the geographical regions encountered, but it also allowed for the discovery of new species like the Black-billed Magpie (*Pica hudsonia*), Common Poorwill (*Phalaenoptilus nuttallii*), Lewis's Woodpecker (*Melanerpes lewis*), Trumpeter Swan (*Cygnus buccinator*) and Sage Grouse (*Centrocercus urophasianus*), depicted opposite in a sketch taken from the journals of the expedition.

Preliminary studies relating to the construction of the transamerican railways would lead to considerable progress in the understanding of the flora and fauna of North America. Baird, working at the new museum in Washington, supervised the publication of the studies (opposite, the title page of volume VI).

The army also played an important role in the discovery of American avifauna. Nearly 40 military physicians collected and observed birds during the second half of the 19th century. Some would not describe the new species themselves, preferring to forward their collections to the museums. The experience gained from exploring the interior of the continent would enable the United States to organize numerous exploratory voyages abroad. Efforts would concentrate on Central and South America, but also the Philippines, which became an American protectorate in 1898 following the Treaty of Paris that put an end to the conflict between the United States and Spain.

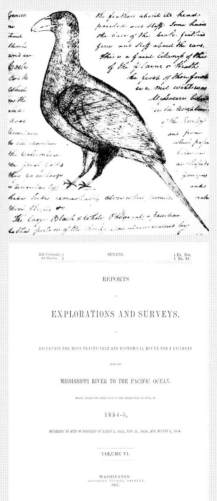

with Audubon, to whom he sent two new species. He would only meet the naturalist–artist four years later during his medical studies in New York. To the great dismay of his circle of friends and colleagues, he abandoned his studies to become a teacher and devote his leisure time to natural history. His scientific work would be recognized in 1848 with the title of doctor *honoris causa*. It was at this time that James Smithson, the illegitimate son of the Duke of Northumberland, donated half a million dollars for the creation of a museum in Washington that would bear his name, the illustrious Smithsonian Institution. Immediately, Baird made use of his contacts

(Audubon, Cassin and other important figures) to seek a position as a curator. In 1850 he started to work as an assistant secretary. A hard worker, he took charge of the organization of the National Museum of Natural History, which is now one of the three largest museums in the world. He directed the museum from 1850 until 1878 and at the same time founded the United States Commission of Fish and Fisheries, which would play a decisive role in the study of marine fauna.

Baird was very adept at taking advantage of all possible opportunities to promote the knowledge of natural history in North America. Beside the organization of numerous expeditions, he placed naturalists in nearly all of the government's research programmes, including geological prospecting, border exploration and the construction of the railways. He would thus empower great ornithologists to launch their careers, including Joel Asaph Allen (1838–1921), future curator of birds and mammals at the American Museum of Natural History in New York, and Robert Ridgway (1850–1929), future curator of birds of the National Museum of Natural History in Washington.

Baird penned *A History of North American Birds* (1874) with Thomas Mayo Brewer (1814–1880) and Robert Ridgway, which would become the standard reference work in American ornithology for two decades. Many birds have been named for him, including Baird's Sandpiper (*Calidris bairdii*), Baird's Trogon (*Trogon bairdii*) and Cozumel Vireo (*Vireo bairdi*).

● The birth of the American Ornithologists' Union (AOU)

During the autumn of 1873, the **Nuttall Ornithological Club** was formed in Cambridge, Massachusetts by young ornithologists William Brewster (1851–1919), Ruthven Deane (1851–1934) and Henry Wetherbee Henshaw (1850–1930). It was dedicated to Thomas Nuttall (see page 136), author of one of the first works on the birds of North America, the *Manual of the Ornithology of the United States and Canada* (1832 and 1834). Despite some early difficulties, the club began to publish its *Bulletin* in 1876. The club gained greater public exposure when it opposed the introduction of the House Sparrow (*Passer domesticus*) to North America. It denounced the aggressiveness of the species in its attacks on young Eastern Bluebirds (*Sialia sialis*), and underlined the fact that it was a rather ineffective insectivore (the primary reason for its introduction).

The members of the club and other ornithologists organized a meeting at the American Museum of Natural History in New York in 1883 that would result in the creation of the **American Ornithologists' Union** (AOU). Among its founding members was **Elliott Coues** (1842–1899). He was an American military doctor who participated in the Borders Commission conducted by Ferdinand Vandeveer Hayden (1829–1887), then joined the United States office of geological and geographical research. He took advantage of his transfers to the Wild West to study its avifauna. In 1872, he published the *Key to North American Birds*. This work would influence a whole generation of American ornithologists. Coues proposed the addition of a third term preceded by the abbreviation *var.* (for variation) for naming variants within species (i.e. subspecies). This system would be developed by Robert Ridgway (1850–1929) and led to the adoption of the trinomial (see page 187) in the nomenclature code used by the AOU in 1886. It was Coues who proposed that this new society start up its own publication, *The Auk*, the first issue of which appeared in January 1884. It remains in publication to this day.

The AOU would become one of the leading organizations in American ornithology. The society held its first conference in 1883 and formed a committee to study the migration of birds, which later expanded to three groups. The first dealt with migration, the second with the geographic distribution of birds and the third with economic ornithology. Cards for noting observations were distributed to those who wished to participate in the survey of migration. It was a huge success. In ten years, 1,000 people contributed, most of them neither ornithologists nor collectors of birds. "The vast majority [of participants] are intelligent farmers, merchants and lighthouse keepers" said one of the men responsible for the committee.

- The other American institutions

The name of **Daniel Giraud Elliot** (1835–1915) is associated with the AOU but also with another American institution: the Field Museum in Chicago. Elliot's feeble health would at times oblige him to interrupt his studies. His first expeditions would take him to the southern United States and the Caribbean. In 1857, he left for Brazil, Europe and the Middle East. He took advantage of his time in London and Paris to visit their museums and form contacts with a number of naturalists. In 1888, he would scour the eastern United Stated in search of bison, but the search would end in vain. He returned with only a collection of bones. In 1894, he took a position as curator in the department of zoology at the Field Museum. In order to improve its collections he organized a number of expeditions to Africa and Asia.

His ornithological works tended to follow the European model of vast monographs on families of birds, rather than the

Elliot Coues (1842–1899) published many important works on the birds and mammals of the American West.

Title page of *A Check List of North American Birds* (1873) by Elliot Coues.

Daniel Giraud Elliot (1835-1915), curator of zoology at the Field Museum in Chicago and cofounder of the AOU. His personal fortune allowed him to publish beautifully illustrated books.

WILD FOWL OF NORTH AMERICA

DANIEL GIRAUD ELLIOT

Cover of *Wild Fowl of North America* (1898) by Daniel Giraud Elliot.

American model, more focused on regional studies. He published *Monograph of the Tetraonidae* (1864-1865), *Monograph of the Pittidae* (1867), *Birds of North America* (1866-1869), *Monograph of the Phasianidae* (1872), *Monograph of the Paradiseidae* (1873), and *Monograph of the Bucerotidae* (1876-1882) among others. He often illustrated his own publications, but also called on the talents of Joseph Wolf (1820-1899) and John Gerrard Keulemans (1842-1912), two illustrators who worked for John Gould. In addition to his publications on birds, he was also interested in mammals.

Elliot completed the specimens collected from his expedition by purchasing historical collections, such as those of Maximilian zu Wied-Neuwied (1782-1867), a German explorer who travelled in North America from 1832 to 1834, which contained a great number of South American species; those of Adolphus Lewis Heermann (1827-1865) which were focused on the American South-west; and a vast collection containing specimens from the world over (including an example of the Great Auk, *Pinguinus impennis*) acquired from the Maison Verreaux (see page 152) in Paris.

To honour his work and his memory, the National Academy of Sciences awards the Daniel Giraud Elliot medal every three to five years to recognize an important publication in zoology or palaeontology.

This overview of American scientific institutions would not be complete without mentioning the museums of New York (which today contains almost a million bird specimens), Washington (nearly 500,000 specimens) and Chicago (more than 400,000 specimens); the Academy of Philadelphia (165,000 specimens); and the zoological museums of the universities of Harvard (more than 300,000 specimens), Michigan (180,000 specimens) and California (180,000 specimens). Of the 15 largest ornithological collections in the world, ten are in North American.

The birth of modern museums in Europe

In Europe, the reorganization of collections at the newly formed Muséum National d'Histoire Naturelle in Paris marked an important step in the emergence of modern museums. Under the direction of Étienne Geoffroy Saint-Hilaire (1772-1844), the collections inherited from Buffon went from 463 specimens (in 1793) to 3,411 (in 1809). Many of the original specimens used by Buffon were lost to the ravages of insects.

If the Jardin du Roi was dominated by Buffon's personality (see page 72), the new museum was quickly controlled by that of Georges Cuvier (see page 183). The Parisian institution was formed by an edict in 1793 and a zoology chair bringing together mammals and birds would be established the following year. This chair would be held by two successive dynasties from 1794 to 1900. The first was that of the Geoffroy Saint-Hilaires, with Étienne (1772–1844), then his son Isidore (1805–1861). The second was that of the Milne-Edwards with Henri (1800–1885), followed by his son Alphonse (1835–1900). Outside of the remarkable studies on bird fossils undertaken by Alphonse Milne-Edwards from 1867 to 1871, these researchers showed little interest in ornithology. It was only in 1900, with the awarding of the chair to Émile Oustalet (1844–1905), that ornithology would occupy a more prominent position. However, it was hard to make up for lost time: despite the receipt of many gifts and rich legacies, the collections of the museum were significantly smaller than those of the British Museum and a number of American and European institutions.

Nonetheless, the orderly organization of the museum's collections was the envy of foreign naturalists. William Kirby (1759–1850) wrote to Alexander Macleay (1767–1848) in 1817 that "every part of the Museum is in beautiful order, systematically arranged, so that every student may in a moment find every object that he wants … I wish the zoological department of the British Museum was in similar order." During the whole of the first half of the 19th century, the British institution lacked space, which would force it to refuse the acquisition of the collections of Sir Ashton Lever and William Bullock (1773–1849).

- The German approach

German ornithology developed more slowly than in France and the United Kingdom, in spite of the work of Philipp Ludwig Statius Müller (1725–1776) of the University of Erlangen, and Johann Friedrich Gmelin (1748–1804) of the University of Göttingen. These two described several species of bird, but above all they were compilers. Charles II August (1746–1795), Duke of Zweibrücken, built up an impressive cabinet of curiosities through the purchase of the collections of Metz-based physician Jean-Baptiste Bécoeur (1718–1777) and of encyclopaedist Pierre Jean Claude Mauduyt de la Varenne (c.1732–1792). His collections were partially destroyed in 1793 during a fire at the ducal palace caused by French Revolutionary troops.

It would only be in 1810 that the first great German museum was founded in Berlin through the combined efforts of Count Hoffmannsegg and his friend Illiger.

Johann Karl Wilhelm Illiger (1775–1813) was the son of a merchant from Brunswick who provided him with a good humanist education and a taste for natural history. He enrolled in the courses of the great entomologist **Johann Christian Ludwig Hellwig** (1743–1831), among others. Impressed with the young man's aptitude, he entrusted the classification of his collection of insects to him, before installing it in his home. Illiger, who suffered from an affliction of the lungs, had to interrupt his medical studies and spent several years studying these collections. Having recovered his health and, on the recommendation of Hellwig, received of a bursary awarded by the Duke of Brunswick, Charles William Ferdinand (1735–1806), Illiger left to study at the University of Helmstedt from 1799 to 1802. Upon his return to Brunswick, he met **Count Johann Centurius von Hoffmannsegg** (1766–1849), who had just returned from an expedition to Portugal and had brought back a splendid collection of plants and insects. Hoffmannsegg, also a student of Hellwig, had undertaken a work on the flora of Portugal in collaboration with naturalist Heinrich Friedrich Link (1769–1851). However, disappointed with the quality of the first illustrations that he received, he decided to settle in Berlin to found his own lithographic printing press.

Illiger lost his patron, the Duke of Brunswick, when he was killed by Napoleonic troops in 1806. Hoffmannsegg called on Illiger to help with the identification of a collection of birds and mammals that he had just received from Brazil around the same time as he was discussing the founding of a museum of natural history associated with the University of Berlin with State Counsellor Wilhelm von Humboldt (1767–1835), brother of the great voyager. He offered his collections and recommended Illiger as a curator. The museum was founded in 1810 in Berlin, Illiger taking on its directorship, but refusing to teach at the university. Martin Hinrich Lichtenstein would take over the teaching of his courses. Illiger devoted himself to sorting through the large number of specimens received, and the new institution could boast 950 species (350 from South America) in 1812, 150 of which were new and described by Illiger. Following Illiger's death in 1813, Lichtenstein would replace him. Hoffmannsegg, who had already lost one of his closest friends the previous year, botanist Carl Ludwig von Willdenow (1765–1812), turned away from natural history and retired to his properties in Saxony. He sold his collection to the Prussian government for the sum of 22,000 thalers.

From 1811, the institution benefited from generous bequests: the collections assembled by Peter Simon Pallas (see page 89) in Siberia as well as the specimens brought back by Carl Heinrich Merck (1761–1799) from the Aleutian Islands, Kamchatka and Alaska. The first collections were substantially enriched by those of Count Hoffmannsegg, which came primarily from America thanks to the efforts of his private collector, Franz Wilhelm Sieber (1789–1844). The museum opened its doors to the public in 1814, already in possession of 2,000 bird specimens.

Illiger, despite his short career, would leave an indelible mark on zoological taxonomy. He sought to establish an accurate terminology for classification and would attempt to redefine the taxonomic ranks proposed by Linnaeus in his *Prodromus systematis mammalium et avium* (1811). He was the first to recommend the translation of Latin terms into German, such as genus into Gattung or species into Art. A purist in the denomination of species, he also put forward the idea of renaming all the taxa whose names were etymologically weak, a suggestion that would receive a lukewarm response, particularly in Germany. Illiger proposed 7 orders, 41 families and 147 genera of birds. In 1812, he presented a paper on climate-based species variation, *Tabellarische Übersicht über die Vertheilung der Vögel über die Erde*, which would only be published in 1816. It was the first publication on ornithological biogeography.

Thanks to the efforts of **Martin Hinrich Lichtenstein** (1780–1857), the Berlin Museum would become one of the most prestigious institutions in Europe. In 1802, following his medical studies at Helmstedt, he became the physician of the Dutch governor in South Africa. On his return in 1806, he brought back a rich collection of specimens (including 340 new species of insect). He published the account of his exploration in the region under the title *Reisen in südlichen Africa* (1811–1812). He took up the zoology chair at the University of Berlin in 1810 and, three years later, the directorship of the museum.

His work as a naturalist was not free from flaws: despite his medical training, he gave little weight to anatomy in the classification of species. Worse, he often sold or exchanged specimens before he even had the chance to study them. Finally, he did not hesitate to describe certain species that had already been described, ignoring the rule of priority, which was largely followed by the scientific community. Despite these weaknesses, he played an essential role in the development of German ornithology by making the Berlin Museum one of the most active centres for research on birds: the ornithological

Martin Hinrich Lichtenstein (1780–1857). After several years in South Africa, he took over the museum of zoology in Berlin, which would become one of the most prestigious institutions of the 19th century.

collections of the museum went from 2,000 specimens in 1814 to 13,760 in 1850. Moreover, he supported the work of Johann Friedrich Naumann (1780-1857) and chose Jean Louis Cabanis (1816-1906) as his assistant.

Lichtenstein was the author of numerous publications, which included several studies on the specimens brought back from scientific expeditions and a monograph on gulls as well as studies on the avifauna of California and the Sandwich Islands. His most important work was published at the end of his life: it consisted of the catalogue of all the birds of Berlin Museum. He would also publish a second list of specimens indicating their prices for those wished to acquire them.

- The Rijksmuseum van Natuurlijke Historie in Leiden

The Netherlands was a strategic hub for collectors in the 18th and 19th century. Its very active ports and network, which included the Dutch East India Company or VOC (founded in 1602), facilitated trade and the distribution of objects brought back from faraway lands, particularly Asia and America. The number of cabinets of curiosities in the country increased regularly, from 28 in the 15th and 16th centuries, to 150 in the 17th century and 540 in the 18th century. A veritable market established itself across Europe: specimens were purchased directly from sailors and explorers, duplicates were traded amongst collectors, and when one of them died, his collection was sold at auction in lots or as a whole. This phenomenon took place throughout Europe, which was afflicted with a genuine infatuation for natural history.

We have already come across the figure of Temminck: Jacob Temminck, treasurer of the VOC, had bought some of the birds assembled by François Levaillant. His son, **Coenraad Jacob Temminck** (1778-1858), also began working for the company and his duties allowed him to make connections with explorers and naturalists. The dissolution of the VOC in 1800 prompted Temminck to dedicate himself entirely to natural history. In 1804, he perfected his knowledge of taxidermy under the instruction of German ornithologist Bernhard Meyer (1767-1836) of Offenbach, who also initiated him to the Linnaean system. Upon his return to the Netherlands, he published *Catalogue systématique du cabinet d'ornithologie et de la collection de Quadrumanes* (1807) in which he detailed the specimens of his private collection, one of the most preeminent in Europe, including several new species of birds. The following year, he started to write the text for the first volume of *Histoire naturelle générale des pigeons* (1808-1811) to accompany the illustrations of Pauline de Courcelles (1781-1851), who became Mrs Knip

Coenraad Jacob Temminck (1778-1858) would have a lasting influence on 19th century ornithology.

in 1808. This work would lead to his being known throughout Europe. They fell out when Mrs Knip decided she would no longer print Temminck's name on the cover of the ninth part of the work.

The first edition of his renowned *Manuel d'ornithologie, ou Tableau systématique des oiseaux qui se trouvent en Europe* was published in 1815 (it would be reprinted in 1820 and revised in 1835 and 1840). In the second edition of his *Manuel,* Temminck lamented the lack of interest in European fauna: "These creatures that surround us seem to have been forgotten by naturalists: we search out the torrid regions and icy poles for new subjects to add to the numerous species already known, by which means we increase the catalogue of nomenclature without any useful scientific goal: sterile acquisitions, that amateurs of curiosities can esteem, that will long be foreign to science." Temminck developed his own classification that included 16 orders and 201 genera.

In 1808, the powers put in place by Napoleon in the Netherlands founded a 'cabinet of the king' (*Kabinet des*

Jacana à crêtes.

The Comb-crested Jacana (*Irediparra gallinacea*), engraving painted by hand by Jean-Gabriel Prêtre for *Nouveau recueil de planches coloriées d'oiseaux* (1820–1839) by Coenraad Jacob Temminck (1778–1858) and Guillaume Michel Jérôme Meiffren Laugier, Baron of Chartrouse (1772–1843).

Konings) that brought together the royal collections. After the departure of the Emperor's brother in 1810, the collection was renamed *'s Lands Kabinet van Natuurlijke Historie*. It was quickly enriched by multiple donations. Among the few people who oversaw it from 1815, Temminck was responsible for the collection of vertebrates. He obtained the support of his childhood friend, Anton Reinhard Falck (1777-1843), Minister of Education, Industry and the Colonies, for a project to create a museum connected to the University of Leiden. This institution was designed to bring together the *'s Lands Kabinet*, the collections of the Academy and those of Temminck. It was founded by royal decree on the 9th of August 1820. Temmick became its first director, a position he would hold for nearly 40 years. The study of collections in Leiden, and also those in the museums of Paris and Vienna, enabled him to describe a number of new species in *Nouveau recueil de planches coloriées d'oiseaux pour servir de suite et de complément aux planches enluminées de Buffon* (1820-1838) which he co-authored with Guillaume Michel Jérôme Meiffren Laugier, Baron of Chartrouse (1772-1843). This luxurious large format book was illustrated with 600 plates depicting 800 species. General consensus seems to be that Temminck was the author of the scientific parts of the work and the descriptions of new species are attributed to him. His last major work was *Fauna Japonica* (1833-1850), written in collaboration with H. Schlegel and based on specimens sent from Japan by Philipp Franz von Siebold (1796-1866), the first westerner to study the fauna and flora of this country.

On his deathbed, he confided that he had achieved his life's ambition: to make Leiden's National Museum of Natural History one of the most distinguished in the world. The collections were improved by virtue of exchanges with other museums and thanks to the efforts of explorers like H.B. von Horstok in South Africa, H.S. Pel in Ghana, and expeditions financed by the Dutch government in South East Asia (see the Natuurkundige Commissie, page 98).

It was **Hermann Schlegel** (1804-1884) who succeeded him in 1858. Born in Saxony, Schlegel studied at Vienna before joining the museum in Leiden in 1825. Like other naturalists before him, he wished to go to Indonesia to gather birds, but Temminck, who was aware of the high rate of mortality, dissuaded him from setting out.

Schlegel's first interests were reptiles and amphibians, before turning to mammals and particularly birds. He would leave behind an important body of ornithological work. He studied the species discovered in the Dutch colonial possessions and participated in the study of the fauna of Japan and

Hermann Schlegel (1804-1884) was particularly interested in geographical variations in species, attributing their origins to divine creation.

Plate take from *De dieren van Nederland. Gewervelde dieren. Vogels* (1860) by Hermann Schlegel.

Madagascar. He encouraged the work of several prominent naturalist artists like Joseph Wolf (1820–1899), Joseph Smit (1836–1929) and John Gerrard Keulemans (1842–1912). Though the Natuurkundige Commissie was dissolved in 1850, the museum benefited from the specimens sent by other voyagers like Heinrich Agathon Bernstein (1828–1865), Dirk J. Hoedt and Carl B.H. von Rosenberg. Schlegel also increasingly turned to purchasing specimens from merchants of natural objects, such as G.A. Frank in Amsterdam and the Verreux Company in Paris (see page 152). It was in this way that he acquired a series of birds gathered by Alfred Russel Wallace (1823–1913) on the island of Seram and in New Guinea.

In addition to his purely ornithological works, *De Vogels van Nederlandsch Indië* (1863–1866), *De Vogels van Nederland* (1854–1858) and *Revue critique des oiseaux d'Europe* (1844), he also published *Traité de fauconnerie* (1844–1853) with Abraham Hendrik Verster van Wulverhorst (1796–1882).

He improved the collections of the museum in Leiden, which he managed quite differently from Temminck, from 1858 until his death. Instead of seeking to assemble the largest number of species, he favoured large series of specimens to demonstrate specific variations. Whereas Temminck would only keep three specimens per species, Schlegel would build a series of at least 20 individuals. The ornithological collections of the museum went from 12,500 specimens at the time of Schlegel's entry in 1858 to 51,000 in 1884.

Knysna Turaco (*Tauraco corythaix*), hand-coloured lithograph by P.W.M. Trap, based on a drawing by Herman Schlegel, from *De Toerako's afgebeeld en beschreven* (1860).

• The British Museum's rise to power

During the 19th century, the number of museums multiplied throughout England (a phenomenon that took place in all European nations) and the country could boast 250 such institutions by the end of the century. The most prosperous museum during the first half of the century was that organized by the Zoological Society of London. Many great zoologists would work there, including John Gould (see page 169), John Gilbert (1812–1845), George Robert Waterhouse (1810–1888) and Louis Fraser (1810–1866). The number of birds conserved surpassed 10,000 specimens. However, the rapid growth of the collections led to storage problems and financial troubles and the collections were completely dispersed in the 1850s.

In the meantime, the naturalist work of British Museum entered into a phase of growth after a long period of inactivity. The ornithological collections of the British Museum consisted of only 1,172 specimens in 1753; 130 years later this number would reach 30,000. This strong growth was possible on account of the work of the Gray brothers, who entered the British Museum to find all the old ornithological collections disintegrating into dust: not a single specimen belonging to

Sloane, Banks, Cook or Latham had survived. A single specimen from the collection of George Montagu had been preserved in mediocre condition.

John Edward Gray (1800–1875) studied medicine and helped his father, botanist and chemist Samuel Frederick Gray (1766–1828) while he was writing of *The Natural Arrangement of British Plants*, published in 1821. Disappointed with botany and botanists after his candidacy to join the Linnaean Society in London was refused, he devoted his entire life to zoology.

As the secretary of the Entomological Society in London, he befriended William Elford Leach (1790–1836), who invited him to assist John George Children (1777–1852), who was engaged in the important work of cataloguing the British Museum's reptile collection. Children, ill informed on zoological matters, gave carte blanche to J.E. Gray who would replace him as the head of the department of zoology in 1840.

J.E. Gray published more than 1,200 publications devoted to various zoological groups, including molluscs, reptiles, amphibians, and mammals. He contributed to the elevation of the collections of the British Museum, which acquired more than a million new specimens under his direction. He would spare no sacrifice to arrive at his ends and did not hesitate to use his own funds to acquire certain collections of interest. He frequently had to fight against colleagues who considered the natural history department the least important of the British Museum.

Gray was one of the first to implement a separation between the specimens bound for exhibition to the public and those intended for scientific study. Even if ornithology was not his primary concern (he left it almost entirely in the care of his brother George), he nonetheless described more than 40 new species.

George Robert Gray (1808–1872) directed the department of ornithology at the British Museum for 40 years, though he had started his career as an entomologist (he published a catalogue of Australian insects in 1833). He devoted the majority of his work to the description of the ornithological collection of the museum that very quickly improved over time. From 1844 to 1848, he published his primary work, *The Genera of Birds*, in three volumes in which he described 11,000 species using 46,000 references. This work would have a considerable influence on several generations of ornithologists.

The two brothers, who dedicated their entire lives to the British Museum, died three years apart, and would not live to see the zoological collections moved into their splendid new home in South Kensington in 1881.

John Edward Gray (1800–1875) was not only a prolific zoologist, but also one of the first philatelists. His carried out a considerable amount of work for the British Museum.

George Robert Gray (1808–1872), brother of John Edward, was in charge of the ornithological collection of the British Museum.

Rufous-necked Hornbill (*Aceros nipalensis*), from *The Genera of Birds* by George Robert Gray (1808–1872).

Title page of one of the catalogues of the collections of the British Museum published by G.R. Gray.

The name Richard Bowdler Sharpe (1847–1909) is synonymous with the rise of the British Museum.

Richard Bowdler Sharpe (1847–1909) would breathe new life into the British Museum. He began very early to develop an interest in animals and collected birds that he stuffed himself. At 15 years of age, he became an editor's apprentice and in 1866, worked for a bookseller located in the vicinity of the British Museum. He began to study kingfishers and conceived the idea for a monograph on the *Alcedinidae*. His work attracted the attention of specialists who frequented the business including Philip Lutley Sclater (see page 122) who would offer him the position of Assistant Librarian at the Zoological Society of London.

Sharpe completed his education by studying the works of Hermann Schlegel (see page 144). He met the naturalist illustrator John Gerrard Keulemans (1842–1912) who would illustrate his monograph, *The Alcedinidae or Family of Kingfishers*, which was published between 1868 and 1871. In 1872, on the recommendation of the director of the department of zoology, Albert C.L. Günther (1830–1914), Sharpe succeeded G.R. Gray to become the director of ornithological collections at just 24 years of age.

The ornithological collection of the British Museum, which had the reputation of being less than attractive, consisted of roughly 30,000 specimens when he entered into the position (of which 10,000 were mounted and exhibited). Sharpe, thanks to the support of influential ornithologists, would completely reorganize it, breaking with the spirit of his predecessor. From 1873, he began to enrich the collection through the acquisition of Alfred Russel Wallace's (1823–1913) esteemed collection of birds from the Malay Archipelago. This would be the first in a long series of acquisitions: the most important collections acquired thanks to Sharpe's initiative were those of Charles Darwin, John Gould, Thomas Stamford Raffles (1781–1826) and Allan Octavian Hume (see page 116). In 1890, Sharpe had acquired 230,000 specimens and, at the time of his death in 1909, this number would surpass half a

Black Sicklebill (*Epimachus fastuosus*), from *Monograph of the Paradiseidae, or Birds of Paradise and Ptilonothynicidae, or Bower-birds* by Richard Bowdler Sharpe (vol. I, London, 1891–1898).

million. In 1972, these were transferred to the museum in Tring (see page 156). Its collection now numbers more than a million skins and is the most important in the world.

Sharpe was not only concerned with the enrichment of his collections (which meant cataloguing, arranging and verifying all newly-acquired specimens), but also with the completion of the work that John Gould's death had left unfinished. Thus, he worked on *The Birds of Asia* (1850–1883), *The Birds of New Guinea* (1875–1888) and *Monograph of the Trochilidae* (1849–1887). He wrote the text for a work on swallows, *Monograph of the Hirundinidae* (1885–1894), published by Claude Wilmott Wyatt (1842–1900).

Two Streaked Bowerbirds (*Amblyornis subalaris*), from *Monograph of the Paradiseidae, or Birds of Paradise, and Ptilonorchynchidae, or Bowerbirds* (1891–1898) by Richard Bowdler Sharpe.

The most striking work by Sharpe was without doubt his *Catalogue of the Birds in the British Museum* which appeared in 27 volumes from 1874 to 1898. This work, which he undertook shortly after being instated in his position, was inspired by Albert Günther's *Catalogue of the Fishes in the British Museum*. He sought to describe all the birds of the world, draw up the list of their synonyms, provide an exact description of their plumage and record their distribution. The scope of the task was immense. Having started in 1874, Sharpe finished the 11th volume in 1890. He admitted that even working almost the entire year, often staying up late, only about 300 species per year could be described in this manner. The work would only be completed after 40 years of work. The director of the British Museum, Günther, authorized him to employ other specialists to lighten his workload. It was thus that Henry Seebohm (1832–1895), Hans Friedrich Gadow (1855–1928), Osbert Salvin (1835–1898), Philip Lutley Sclater (see page 122), George Ernest Shelley (1840–1910), Adelaro Tommaso Paleotti Salvadori (1835–1923), Edward Hargitt (1835–1895) and Ernst Hartert (see page 161) came to be responsible for particular families.

Over the century, the world gained a dense network of museums. Their role would slowly evolve over the 20th century. We will see in the following chapter that three main factors contributed to this: the professionalization of science, the appearance of new disciplines (ethology, ecology and genetics), and the environmental crisis and its consequences for biodiversity.

Collections, collectors and trade

Private natural history collections were very fashionable in the 19th century. Their typical size grew considerably: a collection of 20,000 skins was exceptional around 1800; a century later a collection of 30,000 would become more common. These private collections were sometimes more significant than those of scientific institutions.

The acquisition of new specimens took place through purchases and exchanges, a result of the owners' work or, more often, that of professional collectors. The demand was such that a veritable trade arose. The two most well-known businesses were those of Benjamin Leadbeater (1760–1837) in London and the Verreaux family in Paris, both founded at the beginning of the century.

• The Maison Verreaux and scientific trade

The Maison Verreaux was founded by Pierre-Jacques Verreaux in 1800. Many members of the Verreaux family were closely associated with natural history: the father, Pierre-Jacques Verreaux, managed a store that sold specimens; his sons Jules, Édouard and Alexis would explore faraway lands to supply the family business; and their uncle, Pierre Antoine Delalande (1787–1823), explored America and South Africa for the benefit of the Muséum d'Histoire Naturelle in Paris.

An impressive number of specimens was brought back from Africa by Delalande, who was accompanied by his 11-year-old nephew **Jules Verreaux** (1807–1873): more than 130,000 specimens, most of which were plants, with nearly 300 mammals, more than 2,000 birds and more than 4,000 shells.

Following his return to Paris, Jules Verreaux studied under Georges Cuvier (1769–1832) at the museum. He returned to South Africa in 1825 and met Sir Andrew Smith (1797–1872), who oversaw the recently founded museum in Cape Town. Among the birds that Verreaux sent to Paris was a specimen of a large bird of prey that was purchased by an amateur. René Primevère Lesson (see page 104) was given the opportunity to study it and published his description in *Centurie zoologique* (1830), naming it Verreaux's Eagle (*Aquila verreauxii*). In 1832, a new expedition took **Édouard Verreaux** (1810–1868) and Alexis to Africa. The following year, Édouard left on his own for Sumatra, Java, the Philippines and Indochina. Upon his return to Paris in 1834, he took up the family business. Alexis remained in Africa, where he opened a gunpowder store and continued to send specimens to his brothers. Jules left for Australia. He stayed five years and brought back 11,500 specimens. After this last great expedition, he settled in Paris and helped his brother Édouard.

The Verreaux brothers did not confine themselves to trade. They were also the authors of many publications. Jules studied the specimens sent from China by Père David (see page 118), who discovered the panda and the eponymous deer. He also published a work on the birds of New Caledonia with Oeillet Des Murs (1804–1889). Édouard and Jules also published a book on the geography of Australia.

The Maison Verreaux was reputed for the diversity of its collection, which was one of the most extensive in the world: in the 1850s, one of their advertisements claimed a stock of 3,000 mammals and 40,000 birds. Nonetheless, the labelling of specimens sometimes left much to be desired, as the unfortunate Lord Lilford (1833–1896) discovered. He received a package of eggs collected in southern Russia, whose names

Jules Verreaux (1807–1873) settled in Paris and worked in the family's taxidermy business after travelling in Africa and Australia.

The Chestnut-breasted Coronet (*Boissoneaua matthewsii*), hand-coloured lithograph by L. Bevalet in the second volume of *Histoire naturelle des oiseaux-mouches, ou colibris constituant la famille des Trochilidés* (1874–1877) by Étienne Mulsant (1797–1880) and Édouard Verreaux (1810–1868). Not only a merchant of natural history objects, Verreaux was the author of several scientific publications.

had, it seems, been allocated at random: eagle eggs were labelled as crow eggs, tern eggs labelled as pratincole eggs and the eggs of the Eagle Owl were identified as those of the Black Stork.

• The bird trade in all its forms

The trade of Frenchman **Adolphe Boucard** (1839–1905) consisted not only of providing the specimens to collectors, but also feathers to the milliners of Europe and the United States. We know relatively little about his life. He described, in one of his texts, his attempts to bring back live hummingbirds from California at the age of 12. A a few years later he visited Mexico to collect birds for several naturalists, including Philip Sclater. On his return to France, he started Boucard, Pottier & Co., a business trading birds. The business, which he moved to London in 1890, promised "to stuff animals from hummingbirds to whales for directors of museums and private collectors at a reasonable price". Its catalogues listed the skins of numerous species of birds, including parrots and 319 species of humming-birds. In addition to publications and scientific materials, a stuffed Great Auk (*Pinguinus impennis*) was listed for the price of £50. He published a monthly journal, *The Humming Bird*, from 1891 to 1895. In it, scientific descriptions were printed next to advertisements, which caused many specialists to disregard it.

Clearly combining natural history and business, Boucard published the explicit *Méthode à la portée de tous, pour se faire 2000 à 5000 francs de rente en s'instruisant et en s'amusant, ou Guide pour collecter, préparer et expédier des collections d'histoire* (1871), which explained how to earn a living from the collection of natural history specimens. In 1894 Boucard conceded that "our current mania for collecting has spread to all classes of society, so that nearly everyone is in possession of a gallery of paintings, watercolours, drawings or a good library, an album of stamps ... and others, a collection of humming-birds could be desired by the ladies. It is magnificent, much more varied than a collection of precious stones and costs much less". He also defended the exploitation of hummingbirds for the fashion trade.

Boucard preserved the best specimens that came into his possession. In this way, he built his own collection of over 40,000 specimens that he gave to the Muséum de Paris, as well as 10,000 doubles to the Washington Museum, 8,000 to the museum in Lisbon and as many to the museum in Madrid.

- ● Reverend Tristram, explorer and collector

British Reverend **Henry Baker Tristram** (1822–1906) was one of the great collectors who took great liberties with his fortune. Initiated into the study of birds and insects by his father at a young age, he collected eggs, including those of Jays, Peregrine Falcons, Red Kites, and Hen Harriers. His love for natural history was a family tradition. His great-uncle had corresponded with Gilbert White (see page 85). Orphaned at a young age, he was sent to Durham, where he visited the museum. The curator encouraged him to trap the birds nesting in the roof of the establishment with his bare hands. Tristram was very proud to present him with his first gift, a Eurasian Jackdaw.

He then left to study at Oxford, and graduated in 1844. He spent six months studying the avifauna of Switzerland, then made his way to Geneva to study taxidermy, followed by a year in Italy, where he met ornithologist Paolo Savi (1798–1871) of the University of Pisa.

At 23 years of age, Tristram was ordained a priest. He began to show signs of tuberculosis and was advised to exchange the climate of England for the heat of the tropics. Tristram left for

Reverend Henry Baker Tristram (1822–1906), photographed here dressed for exploration. He sought to identify the birds mentioned in the Bible and determined that the bird described in the Book of Micah as an eagle was in fact a Red-headed Vulture.

Bermuda where he assumed the dual function of military chaplain and governor's secretary. He met several ornithologists, including Colonel Henry Maurice Drummond-Hay (1814–1896) (the first president of the British Ornithologists' Union). His health still declining, he returned to Great Britain after six months in the United States and Canada and settled in Castle Eden in County Durham. His health deteriorated considerably and he was forced to flee the British winter, heading to Algeria in 1854, with no great hopes of returning. Thankfully, the dry North African climate led to an almost miraculous recovery. He travelled the country and made numerous observations. It was during one of his excursions that he discovered the Tristram's Warbler, *Sylvia deserticola*, which he described in 1859. He would recount his travels in *The Great Sahara* in 1860.

In 1855, Tristram made his way to the Eastern Mediterranean and travelled through Greece and Turkey. Three years later, he visited Palestine for the first time. There, he discovered an orange-winged bird of which he sent two examples to his friend Philip Sclater (see page 122) who would dedicate the new species to him as Tristram's Starling (*Onychognathus tristramii*). This voyage and discovery prompted Tristram to plan numerous expeditions in the region, often in risky conditions on account of the high numbers of brigands in the area.

He recorded his numerous and precise observations in several books including *The Land of Israel* (1865), *The Natural History of the Bible* (1867) and *The Survey of Western Palestine: The Fauna and Flora of Palestine* (1884). His descriptions are still used today, allowing one to follow the evolution of the region's environment.

Tristram played a significant role in the development of field ornithology. He wrote in the introduction to his work on the ornithology of North Africa (published in *The Ibis* in 1859)

One of the last specimens of the Labrador Duck (*Camptorhynchus labradorius*) was found in the collection of Reverend Tristram, a gift from his friend Colonel Wedderburn during his stay in Bermuda. The Colonel's widow explained in a 1913 letter that her husband 'often regretted having killed the last Labrador Duck … in 1851. He knew how rare the bird was at the time.'

that "it is impossible to gain an accurate picture of the ornithology of any country without first taking into account its physical and geological characteristics".

A relentless collector, dreading the possibility that his collection might be divided after his death, Tristram offered the museum in Liverpool a set of 17,000 skins, representing 6,000 different species and including 130 biological types. He began a new collection at the age of 74 that would attain the impressive number of 7,000 specimens, representing 3,000 species. This second collection would be acquired by the Academy of Philadelphia. Tristram possessed many specimens of lost species, including the Norfolk Island Kâkâ (*Nestor productus*) and the Labrador Duck (*Camptorhynchus labradorius*).

- Rothschild: the world's greatest collector

The name of **Lord Lionel Walter Rothschild** (1868–1937) is associated with the most vast private collection of the 19th century: he assembled more than 300,000 skins and 200,000 bird eggs, and also, with the help of his brother, entomologist Charles Rothschild (1877–1923), 300,000 beetles and more than 2,250,000 butterflies.

Walter Rothschild (1868–1937), aged 24. He hired Ernst Hartert and Karl Jordan to curate his collections around this time.

The Rothschild family settled at Tring Park, a property situated 50 kilometres from London. At seven years old, Rothschild started to assemble his first collections, which he christened his "museum". In 1886, after being educated by a private tutor, he left to study at the University of Bonn. The following year he enrolled at Magdalene College, Cambridge, where he studied for two years but left without a degree. In 1889, he sent his first collector, Henry Palmer, to the islands of the Pacific to collect birds (his mission lasted three years and he brought back 1,832 birds including 15 which were new to science). The same year, Rothschild began to build his museum in Tring, which would open its doors to the public in 1892.

On the advice of his father, he entered into the family bank, N.M. Rothschild & Sons, at 21 years of age. He would work there for 18 years. According to his niece, entomologist Miriam Rothschild (1908–2005), he detested every day that he spent there. His natural history collection and menagerie were his true passions. He dedicated almost his entire salary (£5,000 a year) to finance artists, improve his collections, and acquire and care for his live animals. In 1893, based on the observation of specimens received from Palmer, he published the first volume of his *Avifauna of Laysan*, an island of the Hawaiian archipelago reputed for its guano harvests. Rothschild was the author of other publications on land-dwelling giant turtles (he raised several species) and butterflies of the *Papilio* and *Charaxes*

Nests of the Laysan
Finch (*Telespiza cantans*),
above, and of the Puaiohi
(*Myadestes palmeri*), below.
Plate taken from *The
Avifauna of Laysan and the
Neighbouring Islands with
a Complete History to Date
of the Birds of the Hawaiian
Possession* (1893–1900) by
Lord Walter Rothschild.

A Laysan Albatross
(*Diomedea* [now *Phoebastria*]
immutablis), illustration from
*The Avifauna of Laysan and
the Neighbouring Islands with
a Complete History to Date
of the Birds of the Hawaiian
Possession* (1893–1900), by
Lord Walter Rothschild.

The Northern Cassowary (*Casuarius unappendiculatus*), lithograph by John Gerrard Keulemans, coloured by hand: a plate from *The Monograph of the Genus Casuarius* (1900), by Lord Walter Rothschild.

CASUARIUS UNIAPPENDICULATUS

genera. In 1897, he commissioned an expedition to the Galapagos Islands, most of which he financed and would receive an honorary doctorate from the University of Giessen in Germany.

In 1899, he published a study on the ornithology of the Galapagos. The same year, he became the administrator of the British Museum and a member of the House of Commons. He continued to publish articles on butterflies and birds, which included a monograph on extinct birds in 1907. In 1908, he stepped down from his position at the family bank. He became a member of the Royal Society in 1911 and inherited the title of Baron from his father three years later. A prominent Zionist, in 1917 he received the letter known as the Balfour Declaration, which stated the British government's interest in establishing a national home for the Jewish people. He retired in 1930.

What distinguished Lord Rothschild from other rich and powerful collectors of his time was his scientific rigour. He surrounded himself with excellent colleagues, Karl Jordan

(1861–1959) in entomology (who would describe alone or with the help of one of the Rothschild brothers no less than 3,400 new species) and Ernst Hartert in ornithology (see page 161). Not only would these two researchers enhance the collection and study the specimens received in Tring, but they were at the forefront of the revival of a more systematic approach. One characteristic of this collection is the large number of specimens of the same species. In general, collectors sought to possess the greatest number of different species (specimens of both sexes and the young to which they would sometimes add eggs and nests), and sold or exchanged doubles. Lord Rothschild did not have this constraint and the material that he assembled has few equals the world over.

Rothschild acquired certain historic collections, including those of Christian Ludwig Brehm (1787–1864), with 9,000 specimens, and the exceptional collection of Gregory Macalister Mathews (1876–1949) with over 45,000 Australian bird skins. Most new specimens were acquired by collectors specially

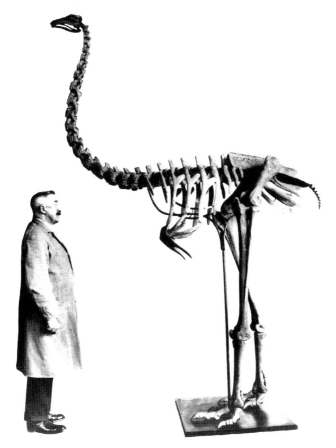

The museum in Tring employed several taxidermists like Fred Young, who is pictured here in front of a reconstituted moa skeleton.

Head of a cassowary, probably a Northern Cassowary (*Casuarius unappendiculatus*), from *The Monograph of the Genus Casuarius* (1900) by Lord Walter Rothschild.

employed by Lord Rothschild, which would allow the exploration of various parts of the world that had, up until then, been largely ignored by naturalists.

Among the ornithologists who trained in Tring were Carl Edward Hellmayr (1878–1944), curator of the Field Museum in Chicago from 1922, Erwin Stresemann (1889–1972), a leading German ornithologist who directed the department of ornithology at the museum in Berlin from 1921 and Ernst Mayr (1904–2005), who played a prominent role in 20th century evolutionary biology.

In 1894, Rothschild launched the museum's journal, *Novitates Zoologicae*. It would enable, in addition to the publication of works on the specimens of the collections, the promotion the use of trinomials for the designation of subspecies (see page 187). Published until 1939, it totalled 40 volumes, roughly 23,000 pages and 600 plates.

In 1932, Lord Rothschild's financial situation was severely compromised in part due to a blackmail demand by one of his old mistresses. He sold most of his collection of bird skins to the American Museum of Natural History in New York and only kept his menagerie, which contained 200 ostriches, cassowaries and rheas, his favourite birds.

Among the employees of Lord Rothschild, **Ernst Hartert** (1859–1933) was one of the most important figures. He started collecting bird eggs at a very young age and learned taxidermy at Kaliningrad. In 1885 and 1886, he and other naturalists participated in an expedition conducted by Eduard Robert Flegel (1855–1886) to Equatorial Africa. From 1887 to 1889 he participated in an expedition to Sumatra, the Malaysian Peninsula and India. Shortly after his return to Germany, he was put in charge of completing a catalogue of the birds of the Senckenberg Museum in Frankfurt am Main, which had 10,000 birds on display. He gained a good reputation by studying this collection, neglected for many years, in less than ten months. The catalogue, which was published as *Katalog der Vogelsammlung im Museum der Senckenbergischen Naturforschen-den Gesellschaft* was the first German-language work to use trinomials in defining subspecies. Hartert then tried to find a position, but the Senckenberg Museum could not hire him for financial reasons and the Berlin Museum was already fully staffed. He tried his luck in Great Britain, but with little success, aside from gaining the friendship of Richard Bowdler Sharpe (see page 148) who, in 1891, entrusted him to write the volume on nightjars and swifts in his *Catalogue of Birds*.

Ernst Hartert (1859–1933) played an important role in ornithology. He advocated the use of trinomials in describing subspecies.

Ernst Hartert met with Lord Rothschild in London, which led to his being selected to carry out a scientific expedition to Venezuela that he had been wanting to undertake for many years. He left for South America in 1892 to collect birds and insects. Political disturbances in Venezuela forced him to change his plans and he instead explored the West Indies. Less than three months after his arrival, he received a telegram from Lord Rothschild requesting his urgent return to England to take up the directorship of his new museum in Tring.

In two decades, the ornithological and entomological collections of this private museum would eclipse all those on the continent and soon rival the British Museum's. Hartert left his duties in 1930. Ernt Mayr (1904–2005), who had gathered birds in New Guinea for the museum in Tring, was for a time approached to fill the position, but the deterioration of Lord Rothschild's fortune put an end to the project. Mayr would nonetheless come to serve as the curator of the collections: hired by the American Museum in New York in 1931, he would supervise the acquisition of the Tring collections.

- ## Fraudulent birds

Collectors' growing demand for rare specimens brought about the appearance of fakes. Made from pieces of birds, they were marketed as entirely unique examples of new species. The quality of the fraudulent birds was remarkable and succeeded in fooling eminent ornithologists. Such was the case of G.R. Gray (see page 147), who was responsible for the birds of the British Museum. In 1841, he acquired a kingfisher from a merchant in Paris that he thought to be new. His examination of the specimen revealed that the wings, legs, body and rump came from four different species. The feathers had been "interlaced artificially to give the impression of a perfect specimen", but the most surprising fact was that the body was that of a species that had not yet been described.

Richard Bowdler Sharpe (see page 148), his successor at the British Museum, described a new species of corvid with the scientific name *Lalage melanothorax*. It would only be seven years later that he discovered the deception: a close examination revealed that body was that of a Black-headed Cuckoo-shrike (*Coracina melanoptera*), the head and the neck that of a Black Drongo (*Dicrurus macrocercus*). Sharpe testified: "The fact that I did not notice this detail earlier is as surprising as the fact that I have shown this specimen to many ornithologist friends, of whom most have an intimate knowledge of the two above-mentioned species, and that none of us detected the fraud. In re-examining the specimen, as I have done countless times before, it is impossible to locate the point at which the birds were joined."

In 1904, Sir Walter Lawry Buller (1838–1906) attempted to sell an adult specimen of an extinct species of small owl that lived in New Zealand: a Laughing Owl, or *Scleloglaux rufifacies*. He offered it for a large sum to Lord Rothschild (see page 156), who examined it closely. He discovered that it consisted of a juvenile to which the rump of a different species belonging to the *Ninox* genus was added. Lord Rothschild would take pleasure in describing this fraudulent assemblage more than once in his writings, to Buller's great humiliation. This specimen has since disappeared, which is regrettable as only 60 Laughing Owl specimens remain.

These examples were not isolated to ornithology and were present in all areas of natural history: fraudulent butterflies, beetles, shells, fossils, etc. were also uncovered. In fact, they

Sir Walter Lawry Buller (1838–1906) was one of the leading specialists on the birds of New Zealand. He sold numerous specimens, including some extinct species, to Lord Rothschild.

were present in any field where the greed of collectors made the creation of such fakes profitable.

The role and place of collectors, and more generally, amateurs in science would markedly evolve over the course of the following century. If the study of nature remained popular, a whole series of factors would change the means by which it was studied. In the field of ornithology, one of the most striking of these would be the abandonment of the gun and the collection of specimens to adopt observational tools that did not result in the death of the birds. Amateur scientists would also face competition from the professionalization of science, particularly new disciplines like ethology, ecology and genetics.

When science became an art, or the golden age of ornithological illustration

We cannot discuss the ornithology of the 19th century without referring to some of the great naturalist artists who created splendid illustrated works, often in a very large format. These deluxe editions with limited print runs (often between 100 and 300 copies) were aimed at a very wealthy audience.

- Audubon: artist, scientist and explorer

Jean Jacques Audubon (1785–1851) was the illegitimate son of a sailor, Captain Jean Audubon. His mother, Jeanne Rabin, was a Creole from Haiti, where his father owned plantations. She died when the young Jean Jacques was barely a year old. He was then adopted by the Audubons and raised by his stepmother, Anne Audubon. In addition to their land in the Caribbean, the family owned a property near Philadelphia and a villa in Nantes. It was here that he was introduced to natural history by Charles Marie d'Orbigny (1770–1856), father of the great naturalist Alcide Dessalines d'Orbigny (1802–1857). It was also at Nantes that Audubon discovered the arts, and learned music and drawing. In 1802, he left for Paris to study, possibly under the painter David (1748–1825). His father, wanting his son to avoid conscription in the Napoleonic Wars, sent him to Philadelphia. He became a naturalized American in 1805 and changed his first names to John James. He married Lucy Bakewell in 1806. From 1807, John James Audubon took part in a series of business dealings, but without much success. These failures drove him from city to city in the hopes of finally succeeding. In 1819, a final failure lost him not only the money inherited from his father, but also that of his wife and his brothers-in-law.

Portrait of John James Audubon (1785-1851), in his explorer's garb.

For several years, Audubon had been earning money on the side by painting portraits of neighbours and friends as well as teaching French. He learned the basics of taxidermy at the museum in Cincinnati and conceived a project to paint all the species of birds living north of Mexico, which he would publish as a book. Wishing to draw from nature, he travelled North America in order to gather and illustrate all the specimens he could find. His expeditions did not prove easy, particularly in the areas that were largely unexplored. Fortunately he was a man of strong constitution and a proficient shooter. His wife Lucy would help enormously: she took a position as a governess in New Orleans and regularly sent him money. However, Audubon's knowledge of ornithology was below par and some accused him of incorporating fantasies into his illustrations.

It was only when he had put together a sufficient collection of illustrations that Audubon began to search for an engraver and printer. His attempts to sell his project having failed in the United States, he left for Great Britain in May 1826. He

Carolina Parakeets (*Conuropsis carolinensis*), from *Birds of America* by John James Audubon. The last individual of this species died at Cincinnati Zoo in 1914.

exhibited his paintings in several cities (including Cambridge, where the audience included Charles Darwin). Audubon met an engraver from Edinburgh, William Home Lizars (1788–1859), who took on the risk of creating engravings of five of the original paintings in a double elephant format (58 cm x 71 cm). His plates depicted birds of different sizes: Wild Turkey (*Meleagris gallopavo*), Yellow-billed Cuckoo (*Coccyzus americanus*), Prothonotary Warbler (*Protonotaria citrea*), Purple Finch (*Carpodacus purpureus*) and Canada Warbler (*Wilsonia canadensis*). In London, Audubon met Robert Havell (1793–1878), a watercolourist, who agreed to colour the plates engraved by Lizars. After having acquired sufficient subscriptions, Audubon was able to print the first instalments of *The Birds of America* in 1827. The print run was only 200 copies, which may seem rather small, but can be explained by the very large format and the fact that each plate, after being printed in black and white, was coloured by hand.

His stay in Britain allowed him to become acquainted with many ornithologists, including Prideaux John Selby (1788–1867), Nicholas Aylward Vigors (1785–1840) and William Swainson (see page 166) who advised him to add text to his plates. With William MacGillivray (1796–1852), he published *Ornithological Biographies* from 1831. Its publication finished in 1839, numbering 3,500 pages.

Until 1839, he split his time between England (to supervise the printing of *Birds of America*) and the United States, before deciding to finally settle in that country. Audubon carried out many expeditions in the American West, still largely unexplored, which helped increase his notoriety: he was in Florida in 1832, in Labrador in 1833, and in Missouri in 1843. He began to lose his sight in 1846 and started to go senile from 1847. His two sons, John Woodhouse (1812–1862) and Victor Gifford Audubon (1809–1860), finished *The Viviparous Quadrupeds of North America* (1845–1846 for the plates and 1846–1854 for the text).

Audubon's illustrations were subjected to criticism: he depicted birds in postures which were anatomically impossible, his search for spectacular settings won over scientific precision, and finally, on occasion he painted birds that he had never seen. Audubon's contribution was a dual one: next to his scientific and artistic work of undeniable importance, he was one of the great figures of American Romanticism. His travels and writings, exalting the love of nature, have greatly contributed to the study and awareness of the country's wilderness. It is therefore not surprising that his name was chosen for the Audubon Society, which is devoted to the protection of birds.

Title page of *Ornithological Biography* by John James Audubon.

ORNITHOLOGICAL BIOGRAPHY.

BIRDS OF THE UNITED STATES OF AMERICA.

THE BIRDS OF AMERICA.

BY JOHN JAMES AUDUBON, F.R.SS. L. & E.

EDINBURGH.

Audubon's ornithological illustrations, which sought to depict birds in as lively a fashion as possible, were sometimes prone to anatomical errors, such as the position of the neck of this Trumpeter Swan (*Cygnus buccinator*).

- The difficult life of an ornithologist at the beginning of the century: William Swainson

The career of **William Swainson** (1789–1855) is linked to the development of lithography. This economical technique allowed the artist to prepare his own engravings without having to make use of the services of a professional engraver. The young Swainson was introduced to natural history through a collection of shells owned by his father, a senior civil servant in the customs service. At 14 years of age, he became an apprentice clerk in the service of his father, and at 18 he landed a position in the British army stationed in the Mediterranean. During the eight years that followed, he was primarily stationed in Sicily and Malta, but he took several trips to Greece and in Italy. He dedicated his leisure time to collecting and drawing plants and animals, primarily insects, shellfish and fish. In 1815, his health failing, he returned to England, and left his duties, going into semi-retirement.

The following year, Swainson decided to undertake a scientific expedition to Brazil that he had to finance himself, not having obtained any funding from museums or learned societies. Arriving in Brazil in 1816, he curtailed his trip when a revolution broke out in the country. In August 1818, he returned to England with a collection of 20,000 insects, a herbarium of 1,200 plants, 760 fish skins and drawings of 120 fish. Furthermore, he had acquired a collection of 760 bird skins, mostly from hummingbirds and toucans.

It was his friend William Elford Leach (1790–1836) who advised him to take an interest in lithography. After three years of intensive work, Swainson began to publish his *Zoological Illustrations* (1820–1823), with 182 plates that he drew, printed and coloured by hand himself.

Swainson's professional life would involve a number of failed initiatives: when Leach left his position at the British Museum for health reasons, Swainson applied for it, but despite the support of prominent naturalists including Georges Cuvier (1769–1832) and Sir William Jackson Hooker (1785–1865), it was John George Children (1777–1852) who got the position. Likewise, a collaborative project with John James Audubon, whom he met in 1828, would not get off the ground.

William Swainson (1789–1855). His life illustrated the difficulties encountered by naturalists trying to earn a living from natural history.

Swainson then decided to devote himself entirely to the illustration of natural history works and increased publication output. He signed all of the illustrations of the second volume of *Fauna Boreali-Americana; or the Zoology of the Northern Parts of British America* (1831) published by John Richardson (1787–1865) based on observations made during his travels to Canada and the Arctic coasts. Swainson published a second series of his *Zoological Illustrations* (1832–1833) as well as a small book on the avifauna of South America, *The Birds of Brazil and Mexico* (1834). He undertook the illustrations of numerous works and periodicals including three volumes of *The Naturalist's Library* by Sir William Jardine (1800–1874): two volumes of *Birds of Western Africa* (1837) and one of *The Natural Arrangement and Relations of the Family of Flycatchers* (1838). He also contributed to 11 volumes of *Cabinet Cyclopaedia* by Dionysius Lardner (1793–1859). It was in this series that he published his *Treatise on the Geography and Classification of Animals* (1835) in which he defended the quinary system of classification developed by William Sharp Macleay (see page 185).

His wife died suddenly and he found himself with five children to care for. Around the same time, he lost his savings in an ill-fated investment in silver mines in Mexico. He decided to sell his collections to the University of Cambridge and acquire land in New Zealand. In 1840, once the last volume of *Cabinet Cyclopaedia* was completed, he left with his new wife. His fortunes would barely improve. His plantation was the target of several attacks by Maoris and he would die 15 years later without having published any more ornithological works. His final mark on the history of ornithology was his influence on a young man, Walter Lawry Buller (1838–1906), who would become an authority on the birds of New Zealand.

The Mountain Trogon (*Trogon mexicanus*), lithograph by William Swainson from *Zoological Illustrations* (second series, 1829–1833).

82.

- John Gould, great illustrator of birds

The number and quality of the works completed by **John Gould** (1804–1881) make him one of the most important illustrators of the 19th century. His father, a gardener in the royal gardens at Windsor Castle, initiated him into his trade. Gould then completed training at Ripley Castle in Yorkshire. He had a long-standing interest in birds and learned the basics of taxidermy. It is on account of this that he was recruited by the newly formed Zoological Society of London to assume the conservation of ornithological collections. John Gould settled in London and married Elizabeth Coxen in 1829, a very talented illustrator who would play a contributing role in her husband's work from the start.

After receiving a collection of birds from the Himalayas which included new species, Gould was able to publish the first instalment of *A Century of Birds Hitherto Unfigured from the Himalaya Mountains*, a large format work (56 cm x 40 cm). Elizabeth drew the illustrations from sketches by her husband. The plates were printed lithographically, then coloured by hand with watercolours. The text was entrusted to Nicholas Aylward Vigors (1785–1840).

In 1832, Gould began the publication of *The Birds of Europe*, for which he wrote the text himself. The work consisted of five volumes, the last of which was published in 1837. Gould also bought lithographs for *Illustrations of the Family of Psittacidae*

John Gould (1804–1881) was not only an artist and scientist, but also a formidable businessman.

The Wompoo Fruit-dove (*Ptilinopus magnificus*), lithograph by H.C. Richter from *The Birds of Australia* by John Gould.

Elizabeth Gould (1804–1841) was her husband's loyal assistant. In this image she holds an Australian species, the Cockatiel (*Nymphicus hollandicus*).

Joseph Wolf (1820–1899) was a German artist who settled in London in 1848. He would become one of John Gould's most regular collaborators.

from Edward Lear, who was facing financial difficulties at the time.

A talented ornithologist, it was Gould who would identify most of the birds brought back by Charles Darwin (see page 109) from his voyage. In 1838, John Gould completed his only exploratory expedition, which took him to Australia. On his return in 1840, he began to publish *The Birds of Australia*, which was completed in 1848 with a seventh volume, and a total of 681 plates. Nearly 200 new species were described in it. This book would be an enormous success despite a high price tag (£115), which limited its distribution to the wealthiest amateurs.

Gould's voyage had a profound effect on him: "I arrived in the country, and found myself surrounded by objects as strange as if I had been transported to another planet." He published only two works on animals other than birds: *A Monograph of the Macropodidae, or Family of Kangaroos* (1841–1842) and *The Mammals of Australia* (1845–1863).

After the death of his wife, shortly after their return from Australia, he would call upon the best illustrators of his time, including Edward Lear, Henry Constantine Richter, Joseph Wolf and William Hart.

Gould was the author of several systematic monographs. He published *A Monograph of the Trochilidae, or Family of Humming Birds* (1849–1861) over five volumes. The size of hummingbirds allowed them to be depicted at their actual size and in order to better render the radiance of these birds' plumage, Gould used an extremely expensive system that consisted of sheets of gold covered with coloured varnish. He described almost all species known today and a number of subspecies then believed to be separate species. Gould would continue to publish a number of works, including *Handbook to the Birds of Australia* (1865), *The Birds of Asia* (1850–1883), *The Birds of Great Britain* (1862–1873) and *The Birds of New Guinea, and the Adjacent Papuan Islands* (1875–1888). Gould's accumulated works were colossal: they totalled more than 3,000 ornithological plates, representing 2,500 species of which many were illustrated for the first time. He was involved with numerous learned societies and was admitted to the Royal Society in 1843, which was a highlight of his career.

Gould's death interrupted the completion of several works, for which the text would be completed by Robert Bowdler Sharpe (see page 148) and the illustrations by William Hart. The large collection of birds that John Gould had built up over the course of his life is now housed primarily in the British Museum; his birds of Australia are in Philadelphia.

ERYTHACUS RUBECULA.

- Edward Lear: poet, ornithologist and artist

Edward Lear (1812–1888) seemed to take pleasure in inventing dark tales of his childhood. He claimed that his Danish-born father was a rich businessman, in possession of many carriages and employing numerous servants, until a serious reversal in his fortunes landed him in prison, leaving his family of 21 children in the depths of despair. However, the reality of the situation was quite different: Edward Lear was in fact the 20th and second-to-last child of a poor family. Of fragile health (he suffered from bouts of epilepsy), his mother entrusted his care to his eldest sister Ann, 22 years his senior. He received very little education and only attended school for a year, at around 11 years of age. A brilliant self-teacher, he had a passion for animals, literature and art. His sisters, who were skilled at drawing, gave him an early introduction to art by having him paint flowers, butterflies and birds. He reproduced and coloured all of the plates of Buffon's encyclopaedia. To earn a living, he taught drawing classes. When he began to frequent art galleries and museums at around 20 years of age, he met Prideaux John Selby (1788–1867), ornithologist and member of the Zoological Society of London. Selby would commission Lear to illustrate his publications. We find his first scientific illustrations in the two final volumes of *Illustrations of British Ornithology* by P.J. Selby and Sir William Jardine (1800–1874), published in 1828.

Edward Lear (1812-1888) at the end of his life. An illustrator and engraver, he published several magnificent illustrated works on birds.

Lear assiduously visited London Zoo, recently opened by the Zoological Society. He was commissioned to illustrate the visitors guide published in 1831 by Edward Turner Bennett (1797–1836). It was produced in two parts, one devoted to quadrupeds and one to birds. Lear then began to work on *Illustrations of the Family of Psittacidae, or Parrots* (1830–1832). This beautiful large-format book was innovative in several ways: it depicted parrots illustrated from living specimens, often at actual size, and it was one of the first deluxe works devoted to a single family. It was possible owing to a process that had just become widespread: lithography. It was not a commercial success, although the 175 copies printed were sold, particularly to eminent naturalists, who thought very highly of it. From 1831, he contributed to several publications, including *A Monograph of the Testudinata* by Thomas Bell (1792–1880), with James Sowerby (1757–1822). His role in Jardine's *Naturalist's Library* series (1833–1843) would expose his work to a broader audience. He then caught the attention of John Gould (see page 169), one of the foremost editors of natural history books in the 19th century, but also a formidable businessman who unfortunately did not hesitate to erase the names of authors from engravings to replace them with his own or that of his wife. His most important contribution to Gould's publications was his work on *The Birds of Europe*. Elizabeth Gould undertook the illustrations of small coloured birds from sketches drawn by her husband, but most of the larger birds, including birds of prey and corvids, were the work of Edward Lear.

In 1832, Lord Stanley (1775–1851) was searching for an illustrator to immortalize the many animals (of which more than 1,200 were birds) that he kept in his menagerie. On the advice of G.R. Gray (see page 147), he observed Edward Lear

Blue-and-yellow Macaw (*Ara ararauna*) (left) and the title page from *Illustrations of the Family of Psittacidae, or Parrots* (1830–1832) by Edward Lear.

drawing birds in the London Zoo and hired him immediately. Lear would complete hundreds of drawings of animals and birds over five years, some of which were reproduced in *Gleanings from the Menagerie and Aviary at Knowsley Hall* (1846). These years were the happiest and most profitable of his life.

His voyage to Rome during the summer of 1837 would mark a turning point. From then on, he would practically never leave the south of Europe and discovered landscape painting. Around 1846, he began to publish his first poetic works based largely on nonsense.

In the 19th century, developments in printing techniques changed the nature of illustrations. The improvements in lithography, which would evolve into chromolithography, lowered the costs of printing works in colour. The trend would accelerate with the appearance of trichromatic and four-

Eagle Owl (*Bubo bubo*), illustration by Edward Lear, from *The Birds of Europe* (1832–1837) by John Gould.

colour photomechanical processes. Traditional drawn illustrations also faced increased competition from photography (see page 193). The publication of deluxe ornithological works would come to an end with close of the century.

The beginnings of bird protection, or how to justify their utility

The German-born **Constantin Wilhelm Lambert Gloger** (1803–1863) played an important early role in bird protection. His name is connected with Gloger's Rule which attempted to establish a link between the colour of animals and the surrounding climate (the birds of humid climates should be darker than those of drier climates). He was also the author of several very popular books such as *Schlesiens Wirbelthier-Fauna* (1833). Thanks to a bursary awarded by the Prussian Ministry of Agriculture, Gloger studied the drafting of a law protecting insectivorous birds. In *Kleine Ermahnung zum Schutze nützlicher Thiere, als naturgemäßer Abwehr von Ungezieferschäden und Mäusefraß* ('Little essay on the protection of useful animals as a natural defence for the damages caused by vermin and mice') in 1858, he condemned the fact that "the reckless destruction of trees has not only put a strain on the birds that feed on the little rodents, but also on the great number of species that do us the great service of destroying worms and insects". Gloger explained that insectivorous birds were at the mercy of birds of prey when required to cross long distances without cover. His publications, like those of other naturalists, would contribute to greater public awareness of the need to protect birds.

- ## The first legislation

For a long time, protective measures had been put in place to protect game: Eleanor of Arborea, a Sardinian princess in the 14th century, protected the falcons of her island; Lima passed laws on the hunt of game from 1555; and the Dutch colony in New England followed suit in 1629. The protection of other birds is more recent.

Austria-Hungary would play a pioneering role in this field. In 1859, Bohemia established a law that banned the destruction, capture and sale of insectivorous birds as well as their nests and eggs. An 1868 Austrian law classed birds in several categories: birds considered harmful (eagles, eagle owls, kites, magpies, etc.); all birds in a second category were protected between 1 September and 31 January; and game birds were treated as a third category. In 1872, Great Britain passed a similar law that protected a number of species of birds from 15 March until 1 August, and in 1880, a new law would aim to limit the trade of bird feathers. In France, a law of 1844 set accepted modes of hunting, aimed primary at net hunting, night hunting and the hunting of so-called transitory birds. From 1874, several

Title page of *Schlesiens Wirbelthier-Fauna* (1833), by Constantin Gloger.

175

19TH CENTURY

attempts were made to protect insectivorous birds, but to little effect.

It was not only the birds of Europe that would be subject to protective measures: several countries in South America voted to protect the hummingbirds so highly sought-after to be sold as finery. These were not the only birds that were hunted to feed the fashion trade in North America and Europe. All exotic or indigenous birds whose feathers could be sold to milliners were intensively hunted.

- • A worldwide organization

Ornithologists gathered on countless occasions to speak on the protection of birds: in Budapest in 1871, 1875 and 1891, Vienna in 1873 and 1884, Rome in 1875, Paris in 1895 and Graz in 1898. Ornithological societies would play an important role in this mobilization. Specialized societies for the protection of birds first appeared locally in Germany and Austria–Hungary towards 1870, then in the United States with the first Audubon Society in New York City (1886), in England with the Royal Society for the Protection of Birds (1889), Germany with the Bund für Vogelschutz (1899), and Switzerland with the Schweizerische Bund für Naturschutz (1909).

One of the most important steps occurred in Vienna with the organization of the first **International Ornithology Congress** in 1884. Organized by the Austrian Ornithological Society and preceded by a grand exposition of stuffed and tame birds, this event sought to encourage the breeding of poultry and ornamental birds, contributing to the widespread diffusion of scientific knowledge and promoting the protection of useful species. Numerous societies and institutions were represented: Jean Louis Cabanis (1816-1906) represented the Deutschen Ornithologischen Gesellschaft, Anton Reichenow (1847-1941) the Berlin Museum, Émile Oustalet (1844-1905) the Société d'Acclimatation. There were also several societies for the protection of animals and groups of breeders. Directed by Gustav Radde (1831-1903), the congress was inaugurated on the 7th of April by Archduke Rudolf (1858-1889). Three issues were on the programme:

- • An international law to protect birds.
- • Research into the origin of the domestic chicken and the improvement of poultry breeding.
- • The establishment of a worldwide network of stations devoted to ornithological observation.

It was the first issue that was deemed the most pressing by the participants of the congress. They echoed the many threats to birds, which were hunted for meat or eggs with devastating

Cover of *Bird Lore*, the Audubon Societies' journal. Its publication was secured by Frank Michler Chapman (1864-1945).

Bird Lore

AN ILLUSTRATED BI-MONTHLY MAGAZINE DEVOTED TO
THE STUDY AND PROTECTION OF BIRDS

Edited by
FRANK M. CHAPMAN

Official Organ of the Audubon Societies

Audubon Department Edited by
MABEL OSGOOD WRIGHT

VOLUME I, 1899

THE MACMILLAN COMPANY
ENGLEWOOD, N. J., AND NEW YORK CITY

effects. Émile Oustalet cited the example of five ships that, around the 1860s, had destroyed, in only three years, 450,000 penguins for their skins and the oil that could be rendered from their flesh, while exploiting two archipelagos in the southern oceans.

The transformation of the landscape in Europe was also condemned, as it had been in North America. The harmful effects of uprooting hedges was criticized by several naturalists, such as Edmond de Sélys Longchamps (1813–1900) who listed the causes of the disappearance of these favourable avian habitats. Certain ornithologists condemned the construction of railroads or the establishment of telegraphic lines. Thus, Coues (see page 137) "evaluated at several hundreds of thousands the losses of passerines, wading birds and palmipeds that perished by running into telegraphic wires".

Some species had already disappeared and others were threatened. Oustalet worried that "therefore, the equilibrium that originally existed has been broken, the natural harmony [has been] fatally disrupted" and brandished the threat of the proliferation of harmful animals because of the disappearance of their predators. The main argument in favour of the protection of birds revolved around their insectivorous diet. Europe had recently been traumatized by the introduction of two ravagers of crops: the Grape Phylloxera (*Daktulosphaira vitifoliae*) in 1863 and the Colorado Beetle (*Leptinotarsa decemlineata*). The native Cockchafer (*Melolontha melolontha*) was also regarded as a pest after 1877. As there existed few chemical means for fighting them, and since these were often only moderately effective, the role of insectivorous birds was of great importance. From the middle of the century, European insectivorous birds were introduced in the United States, Australia and New Zealand to prey on the insects that had been introduced.

The utility thesis for the conservation of insectivorous birds was not without its opponents, particularly among hunters. Paul Eymard (1802–1878) asserted in a brochure titled *La Chasse aux petits oiseaux* (1867), that "they are useless against insects in a plague" and that "birds are indiscriminate and the harm they do to harvests often surpasses the good that they achieve".

The trade of feathers was subject to much criticism, although the trade of hummingbirds found a defender in Adolphe Boucard (see page 153): "A great battle has been undertaken lately against the wearing of the wonderful hummingbird ... One to two million birds are sent each year to Europe ... My opinion is that certain species of birds are so common that their disappearance would require hundreds of years." He also stated that these birds, were they not hunted by humans, would otherwise fall prey to predators.

Most birds of prey were excluded from the first laws protecting species. In 1893, Karl Theodor Liebe (1828–1894) justified their protection through his experiments on their insectivorous diet. Asserting that no animal could be completely useful nor entirely harmful, Liebe suggested that protection should extend to birds of prey in recognition of the role they played in the balance of nature.

Ornithology itself began to establish more respectful practices with regards to nature. In *Field Ornithology* (1874) Coues (see page 137) said: "Never shoot a bird you do not fully intend to preserve, or to utilize in some proper way. Bird-life is too beautiful a thing to destroy to no purpose." However, the attitude of ornithologists was sometimes ambiguous. Alfred Newton (1829–1907) was an active participant in the establishment of the British law of 1880 that sought to limit the excesses of the feather trade. Paradoxically, at the same time, he worked with equal vigour to enrich his collection of eggs. Likewise, the American Ornithologists' Union's committee for the protection of birds, established in 1886, was directed by William Brewster (1851–1919), owner of one of the largest private collections in the United States.

A sign of the interest stirred up by the 1884 Congress, a number of accounts were published for the authorities of different countries, such as that of Émile Oustalet for the ministry of Public Instruction and Fine Arts in France and that of the directors of the American Ornithologists' Union to the Congress in the United States.

This realization would slowly evolve. In 1902, the Paris Convention for the protection of useful birds was passed as a result of the work in Paris from 1895. The majority of European countries signed this convention except Great Britain. The Royal Society for the Protection of Birds' battle against the feather trade did not end until 1921 with the Importation of Plumage Prohibition Act. In the United States, the national association of Audubon Societies federated the different societies for the protection of birds in 1905 and would play a large role in the banning of the sale of feathers.

● The organized observation of birds

German ornithologists gathered in Brunswick in 1875 and decided to form a committee to oversee the periodic observation of German birds. In 1877, this effort culminated in the publication of Jean Louis Cabanis's (1816–1906) *Journal für Ornithologie*. The number of collaborators grew rapidly. In

Austria, the Austrian Ornithological Society created a similar committee under the patronage of Archduke Rodolphe in 1881.

The 1884 colloquium instated a permanent International Ornithological Committee that received the support of many academies of science (including those of Belgium, Argentina, Berlin, Munich, Vienna, Turin, Lisbon, Stockholm, Madrid and Rouen), regional learned societies, religious missions (including the American mission to Beijing and the director of Berlin missions), and meteorological stations. The committee's duty was to pursue the three objectives of the Congress and produce a journal, *Ornis*.

Bird observatories were established in Europe and the United States, as well as in Argentina, Brazil, Chile, India, Japan, Java and New Zealand. The goal of these observatories was to inventory avian populations, including migratory and transitory birds, and also to ascertain which species had become more abundant or more rare. They were also required to provide reasons for changes in the composition of populations. Finally, they were expected to bring attention to cases of hybridization, and provide details on migration, nesting and behaviour.

The French Minister of Public Instruction established a permanent ornithological commission (following the decree of March 15th, 1883), comprising Milne-Edwards of the Muséum and the Institut, Geoffroy Saint-Hilaire, director of the Jardin d'Acclimatation, naturalist and geographer Alfred Grandidier (1836–1921), Léon Vaillant (1834–1914) and Émile Oustalet of the Muséum, and also the director of the meteorological bureau. A form was created on which observers were asked to note details of the local names of birds, their rarity and frequency, their migratory or sedentary behaviour, the direction taken by birds in migration and on the direction of the wind, the state of the atmosphere, the nesting period and the number of nests and eggs, the duration of incubation, the existence of large colonies (notably of Rooks (*Corvus frugilegus*) and herons), unusual species that were not included in the list provided with the questionnaire and so on. This type of survey was established and distributed in all the countries participating in the work of the International Ornithological Committee in order to standardize, organize and promote ornithological observations.

These studies would allow for the publication of lists of species not based on the study of bibliographies or collections, but on a large number of observations, in works such as *Verzeichniss der Vögel Deutschlands* (1885) published by Eugen Ferdinand von Homeyer and consisting of 358 species.

The evolution of classification towards a natural system

Authors from antiquity to the Renaissance had classified birds according to their habitat, diet and a series of anatomical characteristics. The first classification that broke with this model was that of John Ray and Francis Willughby (see page 44). Their *Ornithologiae* was based on the shape of the beak and feet, and the size of the adults. From then on, classifications would be based entirely upon anatomy.

Levaillant was one of the last ornithologists to follow Buffon and reject the Linnaean system. **Louis Jean Pierre Vieillot** (1748–1831) would be the first great classifier after Linnaeus. He first adopted the Linnaean system before proposing his own. We know very little of Vieillot's life. Born at Yvetot in Normandy, he left for Saint-Domingue in the 1780s. On his return to France, he showed Buffon his observations of the birds of the island, but the latter was not interested, having already completed his book. Vieillot returned to Saint-Domingue with his family shortly after the French Revolution. In order to avoid conscription, he took refuge in the United States where he assembled a large number of bird specimens. A yellow fever epidemic claimed his wife and three daughters and he returned alone to France at the end of the century.

Vieillot published *Histoire naturelle et générale des grimper-eaux et des oiseaux de paradis* (1802) with Jean Baptiste Audebert (1759–1800), of which only 200 copies were printed. This work was followed by *Histoire naturelle des plus beaux oiseaux chanteurs de la zone torride* (1806), and *Histoire naturelle des oiseaux de l'Amérique septentrionale* (1808) in which Vieillot began to employ the Linnaean system. *Analyse d'une nouvelle ornithologie élémentaire* (1816), *Ornithologie française* (1823), *Galerie des oiseaux du Cabinet d'histoire naturelle du jardin du Roi* (1821–1826) and *Histoire naturelle des mammifères* (1819–1822) would follow. Vieillot named numerous genera and species, primarily from the Americas.

It must be noted that his *Galerie des Oiseaux*, whose illustrations were prepared by Paul-Louis Oudart (1796–1860), did not win the esteem of the authorities of the Muséum de Paris according to Alfred Newton: "Its draughtsman and author were refused closer access to the specimens required, and had to draw and describe them through the glass as they stood on the shelves of the cases."

Vieillot's classification, comprising five orders (consisting of 57 families and 273 genera), was criticized by Cuvier and Temminck. The latter published a pamphlet called *Observations sur la classification méthode des oiseaux, et remarques sur l'Analyse d'une nouvelle ornithologie élémentaire par L.P. Vieillot* (1817). Temminck thought Vieillot's classification vague and artificial. He added that Vieillot focused too much on the diet of birds and that he ignored more recent works, including his own. Vieillot would respond to these attacks in the chapter on ornithology that he published in the *Nouveau Dictionnaire d'histoire naturelle*. Temminck would quickly respond in the introduction of his own *Manuel*, stating that Vieillot frequently took his scientific names from a variety of works but "without citing the author, or using a less correct synonym name, or by a name that appears to be new, through the deletion of a few letters and an altogether similar composition". Temminck condemned the Frenchman's lack of knowledge of German works. Vieillot would leave no diary of his travels. He died

Plate of South American birds from *Tableau encyclopédique ... Ornithologie* (1790–1823) by Bonnaterre and Vieillot.

What's in a name?

One of Linnaeus's ideas would prove to be a key asset in the promotion of his nomenclature system: the dedication of the Latin names to a person. Translated or adapted, these were sometimes reflected in vernacular names as well. A personal name was used to construct the scientific name of a new species, and colleagues, teachers, students, collectors, friends or members of one's own family could be honoured in this way. Appealing to people's vanity, these dedications would play an important role. Of course, this led to some abuses. The most striking example of this was Embrik Strand (1876-1947), who managed to have several hundred species across the animal kingdom dedicated in his honour, often by his own initiative.

The dedicatee was often the collector of the species, and it was seen as a way of recognizing the courage of those who would have otherwise remained anonymous. Such was the case, for example, of the Great Crested Tern: its scientific name, *Sterna bergii*, conceived by Hinrich Lichtenstein (see page 140) in 1828, commemorating Karl Heinrich Bergius (*c.*1790-1818). Bergius was a young pharmacist who was sent by Lichtenstein to South Africa to collect specimens, and perished during the expedition.

Birds' names were also used to honour a member of the family. Such was the case of the Thekla Lark, *Galerida theklae*, which Christian Ludwig Brehm (see page 120) dedicated to his recently deceased daughter, Thekla (1833-1857).

The history of bird names is often rather curious. Baillon's Crake bears the name of Louis Antoine François Baillon (1778-1851). This bird was described by Pallas (see page 89) in 1776, two years prior to Baillon's birth. Baillon forwarded a lot of birds to Vieillot (see page 180) who discovered among them a small rail to which he gave the scientific name *Rallus bailloni* and the common name Baillon's Crake. In 1804, Jean Hermann (1738-1800) demoted this bird to ranks of a subspecies, *Porzana pusilla intermedia*, but the vernacular name had already come into use in both English and French.

The story of common names is not always so international. Leach's Storm-petrel (*Oceanodroma leucorhoa*) named in honour of William Elford Leach (1790-1836) in English, but vernacular names in other languages do not reflect this. Leach acquired a petrel for the British Museum during an auction in London. Temminck, who was preparing his *Manuel d'ornithologie*, studied and named it *Procellaria Leachii* (the species name written with a capital as convention dictated at the time), a denomination that quickly found its way into the English language (not without some debates as to whether Leach could claim such an honour). Finally, it was discovered that the species had already been described by Vieillot in 1817, but the English name remained.

The scientific name of the Tibetan Snowfinch (*Montifringilla adamsi*) bears the name of its descriptor Andrew Leith Adams (1827-1882), an unusual occurrence. The species was first named by Frederic Moore (1830-1907), but the work was never published and it was Adams who would cite it in one of his own publications, becoming the first to describe the species, and giving the erroneous impression of having dedicated it to himself.

A recent inventory of scientific names showed that most frequent dedicatee was Philip Lutley Sclater (see page 122) with 18 currently valid species, although he undoubtedly shares a certain number of them with his son William Lutley Sclater (1863-1944).

Next are the Germans Finsch (see page 106) and Gustav Hartlaub (1814-1900) with 12 species each, American John Cassin and Briton William Swainson (see page 166) with 10 species, Dutchman Temminck (see page 142) and German Hans von Berlepsch (1850-1915) with nine species and Americans Baird (see page 134) and Robert Ridgway (1850-1929), British Sharpe (see page 148) and Austrian August von Pelzeln (1825-1891) with eight species. To these we could also add the German Gustav Adolf Fischer (1848-1886), but it is most likely that some of these species were in fact dedicated to Johann Fischer von Waldheim (1771-1853).

The author of the greatest number of valid species is, not surprisingly, Linnaeus, with 714 species descriptions. He is followed by Sclater with 429, Vieillot with 295, John Gould with 385, J.F. Gmelin with 356, C.J. Temminck with 351, and Frédéric de Lafresnaye (1783-1861) with 230 species.

greatly impoverished, his passing seemingly unnoticed by his contemporaries.

- ● The role of anatomical criteria in classifications

In *Le Règne animal*, first published in 1816-1817 before being considerably expanded in 1829-1830, **Georges Cuvier** (1769-1832) proposed a classification of birds in six orders based on multiple anatomical criteria such as the shape of the beak, the position of the digits, the structure of the nostrils and the length of the neck in relation to the feet. Though he undertook many dissections of birds, he relied solely on external anatomical criteria in his classification.

Temminck's *Manuel d'ornithologie* (see page 142) was published in 1815, and updated in 1820, 1835 and 1840. It followed the Linnaean system, and was the standard reference for several years. Temminck pleaded for a classification created with care in order to counter the chaotic approach that predominated at the time. He founded 16 orders and 201 genera, almost completely ignoring the concept of families introduced by Illiger (see page 140) in the works of François Marie Daudin (1774-1804). Temminck preferred Cuvier's names to those of Vieillot.

In 1820, **Christian Ludwig Nitzsch** (1782-1837) advocated classifying birds according to the structure of their nasal glands. His system was hardly practical, but it enabled certain groups to be distinguished from each other. In 1829, in *Observationes de avium arteria carotide communi,* he presented a new version based this time on the circulatory system. He separated swifts from swallows, placing the former with hummingbirds. Nitzsch was primarily known for his work on the structure of feathers.

Henri Marie Ducrotay de Blainville (1777-1850), Cuvier's successor at the museum, established a classification based on the study of the sternum. He birds into passerines and non-passerines and in 1821 separated the genus *Menura (lyrebirds)* from the galliformes. One of his students was Ferdinand Joseph L'Herminier (1802-1866), who combined these observations of the sternum with other anatomical criteria in 1827.

The practice of using the sternum as the basis for classification, employed by a number of ornithologists, was contested by **Arnold Adolph Berthold** (1803-1861). This great physiologist, designer of the first endocrinological experiments, examined the role of this bone in classification in his 1831 *Beiträge zur Anatomie, Zootomie und Physiologie,* concluding that it was not a reliable marker for classification.

Title page of *Histoire naturelle des oiseaux de l'Amérique septentrionale* by Vieillot.

183

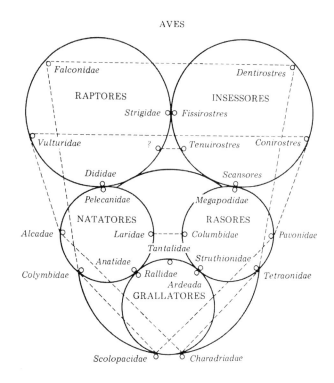

Swainson's classification diagram from 1837. The five orders can be found in the three large circles, one of which is further subdivided into three smaller ones. Each circle includes five families connected to each other by affinities or analogies.

AVES

Falconidae
Dentirostres
RAPTORES
INSESSORES
Strigidae
Fissirostres
Vulturidae
?
Tenuirostres
Conirostres
Dididae
Scansores
Pelecanidae
Megapodidae
NATATORES
RASORES
Alcadae
Laridae
Columbidae
Pavonidae
Tantalidae
Anatidae
Struthionidae
Colymbidae
Rallidae
Tetraonidae
Ardeada
GRALLATORES
Scolopacidae
Charadriadae

To prove the weakness of classifications based on the sternum, he cited the example of the separation of swallows and swifts, which he judged to be mistaken. For Berthold, Linnaean classification could not be improved upon.

Gloger (see page 175) employed multiple anatomical details in his classification of the birds of Europe in *Vollständiges Handbuch der Naturgeschichte der Vögel Europa's* (1834). He divided the order of passerines based on the anatomy of the syrinx (or lower larynx). Johannes Peter Müller (1801–1858) would base the first modern classification of passerines around a similar idea in *Über die bisher unbekannten typischen Verschiedenheiten der Stimmorgane der Passerinen* (1846).

Despite never proposing new classifications, G.R. Gray (see page 147) would play an essential role in improving the taxonomy of birds. He listed the 2,400 genera that were recognized at the time in his *List of the Genera of Birds* (1840–1842) and allocated each species to a genus. He completed this work with a *Handlist* (1869–1871) in which he listed all the names given to genera and species.

- Philosophical classifications

The 19th century saw the publication of several classifications based on philosophical systems. **William Sharp Macleay** (1792–1865) was the author of a classification that has since been forgotten, the quinary system. His father was the naturalist Alexander Macleay (1767–1848). William left Great Britain in 1826 to live first in Cuba before settling in Australia. Between 1819 and 1821, he produced a book on beetles, *Horae Entomologicae*, in which he advanced a theory which connects species to each other through a chain of affinities. Three years later, he extended his system to the whole of the animal kingdom. The quinary system revolved around three principles: first that species form a series of connections based on their affinities, following a linear model; then, relationships can be established between two series of species like the steps of a ladder; finally, any series can be represented by a circle following the affinities of a series of taxa: A with B, B with C, etc., up to E, then E with A closing the circle. This implied that each series had the same number of members, which Macleay set to five, clearly for religious reasons. He believed that this type of consistency in nature had nothing to do with chance, but was the work of God.

Macleay's ideas gained some followers, including **Nicholas Aylward Vigors** (1785–1840), who presented a classification of birds following this method in *Observations on the Natural Affinities that Connect the Orders and Families of Birds* (1824). He created a system based on five orders, each subdivided into five suborders, then each further divided into five subfamilies. The primary utility of his classification was to establish the names of certain families. Moreover, Vigors played an important role in energizing British zoology and contributed to launch of John Gould's career (see page 169). Vigors's theories would find an ardent supporter in the figure of William Swainson (see page 166). This complex system had the advantage of advancing the understanding of certain concepts like homology (a trait can be shared by two species because they received it from a common ancestor) and analogy (an trait can be shared by two different species due to natural selection). The system would be quickly discarded in England, but would continue to exert influence in Germany for several years.

Johann Jakob Kaup (1803–1873) became Temminck's assistant in 1825 before returning to his native city of Darmstadt where he would remain for the remainder of his career at the city's museum. He was a supporter of Lorenz Oken's (1779–1851) theories, which also assigned a particular

Entomologist William Sharp Macleay (1792-1865) was the author of a system of classification based entirely on the number five.

value to the number five. In *Classification der Säugethiere und Vögel* (1844), Kaup based his classification on five anatomical characteristics, the five sensory organs and five parts of the body, but a instead of quinarian circles, he used pentagons. Also inspired by Oken, **Heinrich Gottlieb Ludwig Reichenbach** (1793-1879) produced a series of publications from 1834 to 1863, with the general title *Das Natürlich System der Vögel*. The only whim he permitted into his system was his use of the number four for dividing categories. **Leopold Fitzinger** (1802-1884) followed Kaup's philosophy and proposed the classification of birds into five parallel series following the other classes of vertebrates in *Über das System und die Charakteristik der natürlichen Familien der Vögel* (1856-1865). The emergence of Darwinism would bring an end to these classifications.

- The establishment of the code of nomenclature

British **Hugh Edwin Strickland** (1811-1853) took a very early liking to natural history and geology. He was an opponent of philosophy-based classifications and contributed to the undoing of quinarism in Great Britain. At the beginning of the 1840s, Strickland advocated the establishment of permanent and uniform rules for naturalists the world over. Thanks to his initiative, the British Association for the Advancement of Science appointed a committee of experts who would publish the first code in 1843.

This code prescribed the rejection of synonyms (if a species was validly described twice, it was the oldest name that took precedence) and advocated strict respect for Latin constructions to avoid the introduction of barbarisms in scientific names. Moreover, it established a reference date, 1766, the year of the publication of the 12th edition of Linnaeus's *Systema Naturae*. All the names created before this date were considered invalid.

This code was revised in 1866 and other codes emerged (particularly in Germany). During the 1870s, a number of naturalists called for the 10th edition of Linnaeus's work (1758) to become the starting point for valid nomenclature. In 1895, the first attempts to unify the codes occurred: they would result in the *Code of Zoological Nomenclature* in 1905. After long debates, Brisson's genera, described before 1758 and which therefore should not have been considered valid, would be officially accepted in 1955.

Hugh Edwin Strickland (1811-1853) played an important role in the adoption of a zoological code of nomenclature.

● The appearance of the trinomial in ornithology

Over the course of the second half of the 19th century, ornithologists felt an increasing need to name groups of birds whose appearance was consistent without creating new species.

The formalization of the subspecies designation appeared in the United States with the *American Ornithologists' Union Code* (1886). It was the result of the work of five members of the AOU: Elliott Coues, Joel Asaph Allen, Robert Ridgway, William Brewster and Henry Wetherbee Henshaw, who put forth the idea of adding anothe name to the binomial. For example, the different subspecies of the Golden-fronted Woodpecker were each given a three-part scientific name: *Melanerpes aurifrons aurifrons, M. a. canescens, M. a. dubius*.

In 1884, Coues went to Europe to promote the new rules adopted by the AOU. The meeting, held in London with chief British zoologists in attendance, was not a success and only Henry Seebohm (1832–1895) adhered to the new ideas, the remaining specialists doubting the importance of the trinomial.

German ornithologists, spurred by Gustav Hartlaub (1814–1900), began to tackle this question and a commission was established by the German Ornithological Society in 1890. The following year, it concluded that the difficulties in defining the notion of subspecies were so insurmountable that it was useless to overburden the current system, in which the binomial would suffice in most cases. This conclusion was not satisfactory for Ernst Hartert (see page 161) who, in spite of the commission's conclusion, would define subspecies in his catalogue of the specimens of the Senckenberg Museum. Hartert pursued the promotion of trinomials through the journal *Novitates Zoologicae*, whose publication commenced in 1894.

The heart of the debate was not simply the question of recognizing and naming subspecies. As one of the rare German ornithologists who supported the initiative, Adolph Bernard Meyer (1840–1911), remarked, it was directly tied to accepting of the theory of evolution. The idea was that the isolation of certain populations leads to the progressive appearance of distinct morphological traits. For Meyer, only Hartert's system could account for the designation of these forms, which were on the road to speciation.

Resistance was fierce and it was only at the beginning of the 20th century that a series of ornithologists started to adopt trinomials: the Austrian Viktor von Tschusi zu Schmidhoffen (1847-1924) who directed the periodical *Ornithologisches Jahrbuch*, German Carlo von Erlanger (1872-1904) in the study on birds he had collected in Tunisia, and particularly

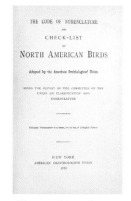

Title page of the *Code of Nomenclature* (1886) adopted by the American Ornithologists' Union.

German Carl Edward Hellmayr (1878-1944), who would become the curator at the Field Museum in Chicago in 1922.

The 19th century drew to a close in a state of complete metamorphosis: petrol became the primary source of energy, cities adopted electricity and the telephone, new modes of transport powered accelerated the movement of people and goods, education became more accessible, particularly to women, and Pasteur's discoveries revolutionalized healthcare. This progress was accompanied by increasingly rapid population growth and urbanization with the consequent destruction of wild habitats. The powerful growth in trade facilitated the spread of live organisms that sometimes became pests in the countries in which they were introduced. The appearance and compounding problems of pollutants would give rise to a major environmental crisis that ornithologists in the 19th century had probably never envisioned.

The 20th century
Ornithology transformed

The 20th century saw the progressive erosion of certain approaches to ornithology, particularly descriptive, anatomical, taxonomic and biogeographical ones. Techniques that did not involve to the killing of birds were preferred to the hunter's trophy specimens of the past. The observation of birds in their natural environment would give rise to numerous studies on their behaviour and contribute to the emergence of a new discipline, ethology. Ringing would become an efficient means of understanding migration. Finally, photography provided an effective substitute to hunting, facilitating an introduction to ornithology and making the publication of illustrated works more commonplace. The metamorphosis was profound and caused ornithology to stand out from the rest of natural history. Ornithology gained widespread popularity and did not suffer from the general loss of interest that affected other disciplines such as botany, entomology and conchology. The threats facing the environment changed over the course of the century. The human population experienced alarming growth, coupled with the gradual abandoning of an agrarian lifestyle and the appearance of immense megalopolises. Following the Second World War, steadily increasing consumption in Western societies was responsible for massive pollution. Ornithologists would be among the most vocal proponents of the protection of the environment. Notwithstanding the richness and complexity of the evolution of ornithology after 1950, this chapter deals primarily with the first half of the 20th century.

Summary

The survival of
traditional illustration

The first great illustrator of birds to emerge in the 20th century is often considered to be the American **Louis Agassiz Fuertes** (1874–1927). Fuertes was discovered by the American ornithological community during an annual meeting of the American Ornithologists' Union (AOU) in 1896. Elliot Coues (see page 137) declared him to be Audubon's successor (see page 163). Frank Michler Chapman (1864–1945) also spoke highly of him: "No one could resist his ready wit, his whole-souled genuineness, his sympathetic consideration, his generosity of thought and deed. Everywhere he made new friends and everywhere he found old ones."

Fuertes' father, professor of technology at Cornell University, named his son in homage to the great American naturalist of Swiss birth, Louis Agassiz (1807–1873). Fuertes took a very early interest in the works of Alexander Wilson (see page 127) and John James Audubon, and attempted his first illustration at 14 years of age. His penchant for natural history was encouraged by his professors at university, zoologist Burt Green Wilder (1841–1925) and botanist Liberty Hyde Bailey

Red-breasted Merganser
(*Mergus serrator*), illustration
by Louis Agassiz Fuertes.

Burrowing Owl (*Athene cunicularia*), illustration by Louis Agassiz Fuertes.

(1858–1954). Following a meeting of the AOU in 1896, he received a number of commissions. At the same time, he continued to study art under the painter Abbott Handerson Thayer (1849–1921).

In 1899, Fuertes participated in a vast expedition to Alaska organized by financier Edward Henry Harriman (1848–1909). It would be the first in a long series of expeditions over the course of which he would paint birds of Canada, Mexico, Colombia, Jamaica, and Ethiopia. In the 1930s, he participated in an advertising campaign for a fizzy drinks company by supplying the illustrations for small collectible cards. On the front, in addition to the advertising message, it identified the bird featured on the back, along with the line "For the good of all, do not destroy the birds."

He completed numerous illustrations for many ornithological works and for *The Auk*, the official publication of the AOU (for which he would illustrate the cover in 1915). He started to teach ornithology at Cornell University in 1923. Only four years later, Fuertes was killed when his car was crushed by a train.

Illustration by Fuertes for the cover of *The Auk* (1915), the journal of the American Ornithologists' Union, which would become its official logo.

Most European ornithologists recognize the name of Peterson and the famous guides associated with it. The parents of **Roger Tory Peterson** (1908–1996) were immigrants from Europe who settled in the United States at the beginning of the 20th century. Taking an early interest in nature, the young Peterson saved the money he earned delivering newspapers in order to purchase Chester Albert Reed's (1876–1912) ornithological guide, a pair of binoculars and a camera. Forced to work from the age of 17, he started as a decorator for a furniture manufacturer. His boss, recognizing the young man's talent, encouraged him to enrol in art classes.

He attended a demonstration organized by the American Ornitologists' Union in 1925: there, he met the great ornithologists of his time, like Ludlow Griscom (1890–1959) and the naturalist-artist Louis Agassiz Fuertes (1874–1927). Both men encouraged him to persevere in ornithology. Two years later, he entered the Art Students League before being admitted to the National Academy of Design, a famous art school in New York. To support himself while studying, he first worked as a decorator for another furniture manufacturer then became a teacher. At the beginning of the 1930s, he worked for the Rivers School, a private school attended by Boston's high society.

At this time, the identification of birds was essentially accomplished by taking measurements after they had been captured. Peterson envisioned a guide that would enable the identification of species from a distance. In it, the essential details for species identification would be indicated through illustrations and a precise system of labels.

He worked on it most nights for three years. However, the current economic situation in the country meant that publishers were relunctant to take it on. After having tried his luck with several publishers, Houghton Mifflin in New York expressed its interest, but remained doubtful of the profitability of the project. Peterson then proposed to give up his royalties on the first 1,000 copies sold. *A Field Guide to the Birds. Giving Field Marks of All Species Found in Eastern North America* was published in 1934 and would immediately be a runaway success: the initial 2,000 copies were sold in 15 days. The book, with its corresponding edition on the birds of the west of the continent, would see 47 reprints and three major updates. The publisher would develop similar guides for other areas of natural history and would publish more than 50 titles during half a century.

In 1934, Peterson was recruited by the National Audubon Society as its Educational and Artistic Director. He would publish a number of popular books for both adults and children. He also regularly contributed self-illustrated articles to *Life* magazine. After the Second World War, he travelled the world and undertook the adaptation of his guide, often with the help of other naturalists. In 1938, he co-authored *A Field Guide to the Birds of Britain and Europe* with Guy Mountfort (1905–2003) and Philip Arthur Dominic Hollom (1912–), which was prefaced by Sir Julian Huxley (see page 200). During the 1960s, Peterson got involved in the environmental protection movement and produced a campaign to ban the use of DDT. From then onwards, Peterson would travel the world to observe birds and would leave behind more 150,000 slides. He received a host of prizes and awards throughout his lifetime, of which no fewer than 23 were honorary doctorates.

Ornithological photography, between illustration and protection

The late 19th century saw the emergence of scientific photography due to three main factors. The most important was rapid technical innovation: faster shutter speeds and the improvement of light-sensitive surfaces rendered them lighter and easier to transport, and optics and lighting became more efficient. The second factor was essential to ensuring the popularity of this new technique: advancements in the printing of photographs made it possible to publish illustrated works at reasonable prices, which led to profound changes in the way books on nature were published. Finally, these new images attracted the public's attention and fascination, as much if not more than the beauty of nature itself.

A byproduct of the growing interest in scientific photography, the Zoological Photography Club was created in 1899. Then, in 1912, in London, the first exposition of bird photography was held. In 1904, a special camera adapted to animal photography was marketed by **Oliver Gregory Pike** (1877–1963), the Bird-Land Camera. He popularized his camera in a book published in 1902, *In Bird-Land with Field-Glass and Camera.*

The photographs were not simply a way to replicate the work of traditional illustrators, but ushered in a new way of observing and capturing images of nature. Equipped with a device he described as a zoopraxiscope, British-born American **Eadweard Muybridge** (1830–1904) attempted to capture the movement of the body through a series of shots taken in bursts.

Cover of one of Oliver Gregory Pike's books, *Adventures in Bird-Land* (1907). He can be seen hanging from the cliff with his camera and tripod on his back.

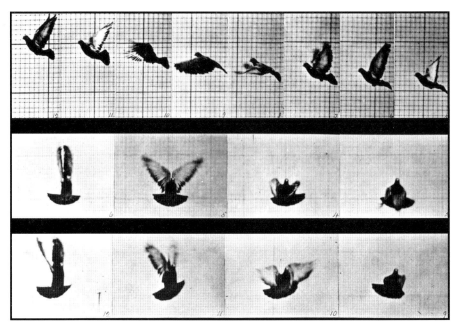

Taken from plate no. 755 of *Animal Locomotion* (1884–1885) by Eadweard Muybridge, breaking down the flight of a pigeon.

Cover of one of the books by the Kearton brothers, *Wild Life at Home, How to Study and Photograph It* (1898).

This work would spark the interest of Frenchman **Étienne-Jules Marey** (1830–1904) and German **Ottomar Anschütz** (1846–1907). The first would develop chronophotography and the second the electrotachyscope. These two devices were designed to capture movement through burst photography: for the first time, we would be able to capture movements too fast for the human eye to be able to detect, freezing the flight of a bird or the landing of a duck. Their work was closely tied to the birth of cinema, as they would devote themselves to reproducing the animated sequences from their series of static images.

Field observation also relied on the spread of animal photography. However, this required the development of techniques of framing, camouflage, remote shutters, and the use of lures or traps.

Two British brothers, **Richard Kearton** (1862–1928) and **Cherry Kearton** (1871–1940), would become famous for their bird photography. They would publish numerous works and would gain renown through the conferences they held in Great Britain, the United States and Australia. Their *Wild Life at Home: How to Study and Photograph It* (1898) was a photographic guide aimed at amateurs. They used their own photographs in many of their works, which totalled about 40, including *Our Rare British Breeding Birds: Their Nests, Eggs, and Summer Haunts* (1899), *Our Bird Friends* (1900), *British Birds' Nests, How, Where, and When to Find and Identify Them* (1908), and *With Nature & a Camera; Being the Adventures and*

Observations of a Field Naturalist & an Animal Photographer (1911). The public infatuation with animal photography made this a rather profitable activity, at least for the best-known photographers.

If the Kearton brothers lived for photography, most other photographers practised this activity on the margins of their career, which generally was related to nature: Francis Hobart Herrick (1858-1940), author of *The Home Life of Wild Birds; a New Method of the Study and Photography of Birds* (1902, reprinted in 1905), taught biology in Cleveland; Frank Michler Chapman (1864-1945), author of *Bird Studies with a Camera: with Introductory Chapters on the Outfit and Methods of the Bird Photographer* (1903), worked for the American Museum of Natural History in New York; Bengt Berg (1885-1967), author of *Mein Freund, der Regenpfeifer* (1912), worked for the Museum of Natural History in Bonn.

The vast number of egg collectors at the beginning of the century explains why many of these new books were devoted to nests and eggs. The advent of animal photography would contribute to changes in the practice of ornithology, as Australian naturalist and journalist Charles Leslie Barrett (1879-1959) pointed out: "Until 1906, I was an ardent collector of eggs, but a conference by Kearton in my school convinced me to put an end to the boring and blowing of shells and to take up the photography of birds, which is what I have been doing for the last 40 years." In fact, these pioneers of animal photography actively rejected traditional hunting and the promotion of the protection of species and their habitats. **Herbert Keightley Job** (1864-1933), a professional ornithologist for the State of Connecticut and member of the Audubon Society, wrote in his preface to *Among the Water-fowl* (1902): "Far be it from me to deny that there are legitimate uses for the dead bird. But owing to relentless, short-sighted slaughter, hiterto carried on, it is coming to be a question of birds or no birds ... If the destruction of life can be minimized by the finding of some satisfactory substitutue for the gun, no one will be the loser."

Photography played a significant early role in the evolution of modern ornithological practice, namely by condemning its connection with hunting. The photographic trophy replaced the hunter's trophy, and it was no longer necessary to kill birds.

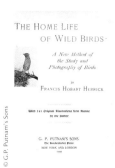

Title page of *The Home Life of Wild Birds* (1902) by Francis Hobart Herrick.

Half title of *Wild Wings* (1905) by Herbert Keightley Job.

Otto Herman (1835-1914) was an important player in the organization of ornithological observations. Above, the title page of his *Method for Ornithophaenology* (1905) which sought to help organize the ever increasing amounts of data amassed on the migration of birds following the work of the first Ornithological Congress in Vienna.

The distribution of species began to be mapped very early. Hungary had a head start on most countries and the first maps were published Otto Herman in 1905.

The marking of birds and the study of migrations

Over the course of the 19th century, several experiments had been undertaken in the individual marking of birds: in 1803-04, Audubon (see page 137) had placed silver rings on the legs of Eastern Phoebes (*Sayornis phoebe*); around 1830, the Dutch Baron van der Heyden had marked the letter H on a brass band placed around the necks of a brood of ducks; later, around 1890, Lord William Percy used copper bands on Eurasian Woodcocks (*Scolopax rusticola*). These isolated experiments, lacking any conclusive results, were taken up by the Dane **Hans Christian Cornelius Mortensen** (1856-1921). He first used zinc, but the rings were too heavy for the birds and injured them. Mortensen then changed the material, taking advantage of improvements in the production of aluminium in the 1890s, which had lowered its cost. In 1899, he ringed 165 Starlings (*Sturnus vulgaris*). Two would be recovered: one from the Netherlands, the other from Norway.

The German **Johannes Thienemann** (1863-1938) took up these experiments in 1903 at an ornithological observation station, the first of its kind, that he had founded two years earlier at Rossiten in East Prussia (modern day Rybachy in Russia). The station, Vogelwarte Rossiten, remains in operation under the direction of the Max Planck Institute. Similar experiments would soon be carried out in other regions of Germany, and also in Hungary and Great Britain. In 1909, the first wide scale ringing operations were organized in the United States and Great Britain. The first national ringing initiative was organized in the United States in the 1920s by the biological research office of the Department of Agriculture. For the first time, traps were used to capture birds, which enabled for greater numbers to be ringed. In 1927, this system was extended to 17 countries. France's first network to study the migrations of birds was put in place in 1930 under the impetus of Édouard Bourdelle (1876-1960) of the Muséum National d'Histoire Naturelle in Paris.

The summary provided by Hungarian **Ottó Herman** (1835-1914), during the fourth Ornithological Congress in London in 1905, epitomized the hesitation that characterized scientists' view on migration. Herman questioned the precision and consistency of migratory routes, and hesitated on the question of whether these were instinctive or not. More data and advancements in physiology would be required before these questions could be answered.

Bird ringing would contribute to the understanding of many aspects of species biology. American **Margaret Morse Nice** (1883–1974), of the Clark University (Massachusetts), was the first to evaluate the rate of mortality in a population of Song Sparrows (*Melospiza melodia*). In 1927, she began to study the territorial behaviour of this species and employed different colour rings to distinguish individual birds at a distance. In 1938, she would make her way to Europe to study with Konrad Lorenz.

In much the same way, British David Lambert Lack (1910–1973) analysed the mortality of the European Robin (*Erithacus rubecula*) and achieved results similar to those of Nice. However, their analyses were met with resistance in the scientific community, because the rates that they obtained were found to be lower than expected. Lack worked with radar technology during the Second World War and he pioneered the concept of using this new technology to study migration, particularly nocturnal migration.

Knowledge of migrations advanced at a dazzling pace. In 1926, Swiss Maurice Boubier, in his work *Les Oiseaux, l'Ornithologie et ses bases scientifiques*, presented a map of the primary migratory pathways. However, the biological and ethological mechanisms by which birds located their route would continue to be debated.

Ethology: the birth of a new discipline

The study of animal behaviour was, for a very long time, one of the pillars of natural history. However, it was in the 20th century that ethology became an autonomous scientific discipline. Many factors can explain this phenomenon: the theory of evolution provided a conceptual framework within which to interpret and compare behaviours; the West developed increasing sensitivity towards nature and greater respect for life, which would lead to the elimination of guns in the practices of collectors; as urban development proceeded, people were increasingly attracted to the observation of nature; and finally, the professionalization of science would facilitate the emergence of new disciplines. It would be echoing the words of many of the forerunners of modern ethology to highlight the enduring quality of many of these themes.

We have already discussed important figures such as Frederick II of Hohenstaufen, Pernau and Zorn, who, in studying birds in their natural environment and through breeding them, were at the root of many discoveries about behaviour. Unfortunately, however, their research did not attract a following. The great precursor of French ethology was Charles Georges Leroy (1723–1789). This officer of the royal hunt published, in letter form, texts in which he discussed the behaviour as well as the sensibilities of animals. One of the major themes of the Enlightenment was the opposition of the notion of the animal-machine: animals, like humans, were equipped with reason.

Several authors would place behaviour into an evolutionary context: Erasmus Darwin (1731–1802), Charles's grandfather, demonstrated in his *Zoonomia* that certain behaviours ascribed to instinct in reality required a period of education; Jean Louis Cabanis (1816–1906) believed that instincts were not fixed and that they responded to the demands of the environment; but it was Jean-Baptiste de Lamarck (1744–1829) who was the first to place instinct within a truly evolutionary perpective. For Larmarck, a new behaviour, performed repeatedly, became hereditary and was transmitted to future generations. Charles Darwin would take an interest in instinct and its role in evolution. Darwin paid particular attention to comparative study of behaviours as can be seen in his book *The Expression of the Emotions in Man and Animals* from 1872. His followers, like George John Romanes (1848–1894) in Great Britain, would contribute to the birth of comparative psychology.

The term ethology was first penned in English by John Stuart Mill (1806–1873) and passed into French 1854 thanks to the work of Étienne Geoffroy Saint-Hilaire (1772–1844). The word's meaning long remained vague, designating the study of animals in their natural environment. In other words, certain authors of the 19th century used ethology to define what we would consider today as ecology.

The notion of tropism was developed by Jacques Loeb (1859–1924), a German biologist who settled in the United States, from experiments on marine invertebrates. His work would be built upon by **Wallace Craig** (1876–1954) in his research on the behaviour of pigeons. Craig was interested in the appetite instinct: the animal will display signs of agitation until this instinct is satisfied. He demonstrated the existence of an aversion principle that worked in a similar way: the animal would remain agitated until a certain stimulus ended. He studied the relationship between behaviour and learning, be it in the construction of a nest or the way in which animals drink. Craig also carried out pioneering studies on bird song and asserted that the vocal imitation ability of certain birds was a development of models of instinctive song. He studied the Eastern Wood Pewee (*Contopus virens*) in great detail and confirmed the observations of other scientists on the aesthetic improvement in song through repeated practice, an improvement that was dependent on the individual's capabilities.

British **Edmund Selous** (1857–1934) was known for his observation of sexual and social behaviour. An original researcher, he refused to enter into contact with other ornithologists and get to know their work. He wanted to come to his own conclusions based on his observations. He also protested against collectors: "For myself, I must confess that I once belonged to this great, poor army of killers, though, happily, a bad shot. But now that I have watched birds closely, the killing of them seems to me as something monstrous and horrible." This refusal to kill was not a completely new approach. Nearly a century earlier British John Freeman Milward Dovaston (1782–1854) refused to kill the birds that he observed, which would not prevent him from making numerous observations on the territory of Robins. In 1924, Jacques Delamain (1874–1953) wrote, regarding Selous, that "his contribution to ornithology, strengthened by its literary value, the mark of love and respect for nature and its creatures, remains that of a master field naturalist, perceptive and confident in his observations, bold and stimulating in his interpretations, and one who passionately endeavoured to penetrate the mystery of life". Selous would undertake very precise observations and would be active as a promoter of a different kind of ornithology: one that did not involve the killing of birds, but the passive observation of them. In spite of the disdain with which the scientific community treated him, he nonetheless inspired several researchers, including Eliot Howard, F.B. Kirkman, Julian Huxley and Edward Armstrong.

Henry Eliot Howard (1873–1940) was interested in the avian territorial systems and published *The British Warblers* (1907–1914) and *Territory in Bird Life* (1920, reprinted in 1948). Howard was a businessman from Worcestershire who devoted his free time to the observation of birds. He published a comparative study on the Reed Bunting (*Emberiza schoeniclus*) and the Yellowhammer (*Emberiza citrinella*) in which he meticulously described their behaviour, the establishing of song territories, courtship and nest building. This book was considered a model for the study of bird behaviour. During the same period, another amateur, **Edward Allworthy Armstrong** (1900–1978) devoted himself to the study of the behaviour of Wrens (*Troglodytes troglodytes*), which would be the subject of his monograph, *The Wren* (1955). Inspired by the work of his fellow Briton, **Sir Julian Huxley** (1887–1975) who published a series of observations on the courtship displays of grebes in

BIRD WATCHING

BY
EDMUND SELOUS

LONDON
J. M. DENT & CO. ALDINE HOUSE
29 & 30 BEDFORD STREET, W.C.
1901

Title page of *Bird Watching* (1901) by Edmund Selous.

1914. Huxley repositioned his interpretation of this behaviour to fit Darwin's theory of natural selection.

The work of **Oskar Heinroth** (1871-1945) would inspire a number of researchers in Central Europe. First, he took up Frenchman Alfred Giard's (1846-1908) call to make ethology an autonomous science, clearly distinguished from ecology. Then, with the help of his wife, Katharina Heinroth (1897-1989), he published an important work on the birds of Central Europe, *Die Vigel Mitteleuropas* (1924-1934). Building on Douglas Spalding's (1840-1877) research, he popularized the concept of imprinting (Prägung). Konrad Lorenz and Nikolaas Tinbergen would recognize Heinroth's influence in their work.

It was in the 1920s, following the work of British biologist Francis Hugh Adam Marshall (1878-1949), that scientists began to shed light on the role of internal (physiological) and external (perceptual) factors in the triggering of reproductive cycles. Eliot Howard took into account the impact of this research in his *Introduction to Bird Behaviour* (1929), *The Nature of the Bird's World* (1935) and *A Waterhen's World* (1940).

In 1936, Huxley and Frederick Bernulf Kirkman (1869-1945), another amateur author who made important observations on colonies of Black-headed Gulls (*Larus ridibundus*), founded the Institute for the Study of Animal Behaviour.

Jakob von Uexküll (1864-1944) occupied an important place in the history of ethology. Uexküll discovered zoology at Tartu in Estonia before taking an interest in physiology at Heidelberg. He lost his personal fortune during the First World War and was forced to integrate into the German university system. His time at the zoological station in Naples, one of the great centres for physiological research in Europe, exposed him to the world of invertebrates. Uexküll sought to analyse the perception of the environment from the point of view of animals, which he referred to as his "experienced worlds", because "everything a subject perceives becomes its world of perception, and all that it does, its world of action. The worlds of action and perception together form a closed whole, the milieu, the experienced world". Uexküll's work would only be translated into the French *Mondes animaux et monde humain* 21 years after his death.

German Frisch, Austrian Lorenz and Dutch Tinbergen were awarded the Nobel Prize in 1973 for their work on behaviour. Karl von Frisch (1886-1982) was primarily known for his work on the language of bees. **Konrad Lorenz** (1903-1989) studied the behaviour of birds that he reared from eggs hatched in an incubator. He was able to demonstrate that certain behaviours are genetically programmed and that the experiences that shaped the early lives of animals had a decisive impact on their future development. Thus, ducks and geese follow the first moving object that they see after hatching, a phenomenon that would be coined imprinting. Studies on corvids, undertaken from 1937, led him to develop the notion of social triggers: a stimulus triggers a behaviour in a member of a group of animals which results in similar reactions in other members of the group. Lorenz's observations on birds brought him to the conclusion that instinctive behaviours are always carried out to completion, even if the animal receives no response to its behaviour from the environment. Beginning in the 1960s, Lorenz turned to the study of human behaviour, which would give rise to intense controversy (particularly when he asserted that aggressive behaviour is an instinctual trait in human beings).

Konrad Lorenz (1903–1989) photographed here with one of his famous geese.

Nikolaas Tinbergen (1907–1988) built on these experiments to measure the power of different stimuli. He noted that exaggerated artificial stimuli produced a more marked behaviour than those occuring under normal conditions. Following his dissertation on the behaviour of the digger wasp of the *Sphecidae* family, in 1931, he studied two birds of Greenland: the Red-necked Phalarope (*Phalaropus lobatus*) and the Snow Bunting (*Plectrophenax nivalis*). His research was interrupted during the war when he was imprisoned by the Nazis after having protested against the treatment and exclusion of Jews. Disappointed by the lack of interest for ethology in Europe, he left for Oxford in 1949 where he would remain for the rest of his career. One of his most important contributions was a series of experiments on feeding behaviour in Herring Gull (*Larus argentatus*) nestlings. These nestlings tap their parents' beaks to get food. Tinbergen hypothesized that it was the red spot at the end of the beak that provided the stimulus. He built models of gull heads, some with the red spot, others without the spot. His results confirmed that the nestlings only responded when the red spot was present. His postwar research dealt with signals used by different species of gulls, and his observations allowed him to piece together the evolution of these signals from one species to another.

Nikolaas Tinbergen's (1907–1988) studies on marine birds have remained famous.

Finally, Tinbergen was one of the pioneers in the study of certain behaviours (nesting in colonies and the placement of nests) as a means of guarding against predators.

Beyond the progress of science, the work of these ethologists would generate a huge amount of public interest: the works of Uexküll, Lorenz and Tinbergen experienced record exposure in the form of animal documentaries that popularized some of their theories. Like ecology, ethology would contribute to greater awareness of the complexity and richness of the living world. In some ways, this success rekindled a traditional notion of natural history, that of Buffon, very far removed from nomenclature and the description of species.

From Darwinian theory to the genetic revolution

In 1893, it was widely believed that German scientist **Hans Friedrich Gadow's** (1855-1928) work on bird classification in *Klassen und Ordnungen des Thier-Reichs de Heinrich Georg Bronn* (1800-1862) formed the basis for modern classification. Gadow was profoundly influenced by the important work of German anatomist Fürbringer (1846-1920) in *Untersuchungen zur Morphologie und Systematik der Vögel* (1888).

All of the work that would follow would be influenced by the work of these two scientists. Such was the case of American Alexander Wetmore (1886-1978), who, in 1930, began to publish updates to Richard Bowdler Sharpe's *Hand-list*, to American James Lee Peters's (1889-1952) *Check-list of Birds of the World* (7 volumes, 1931-1952), and to German-born American Ernst Mayr (1904-2005) and American Dean Amadon's (1912-2003) *A Review of the Dicaeidae*. The German Erwin Stresemann (1889-1972) noted that "as far as the problem of the relationship between the orders of birds, some remarkable researchers have worked on their classification in vain, and little hope remains for impressive breakthroughs... The science ends where comparative morphology, comparative physiology, and comparative ethology failed after nearly 200 years of work". For Americans Charles Gald Sibley (1917-1998) and Jon Edward Ahlquist, who published *Phylogeny and Classification of Birds: A Study in Molecular Evolution* (1990), it was molecular biology that would provide the solution. The technique of DNA hybridization was first developed by Sibley and Ahlquist and later applied to avian taxonomy by Sibley and Burt Monroe (1930-1994). They published *Distribution and Taxonomy of Birds of the World* (1990) in which they defined 2,057 genera and 9,672 species. The drastic changes ushered in by their work would draw numerous criticisms and were not adopted by all ornithologists (not least because it would involve changing habits).

Beyond questions of classification, the development of genetics had an unexpected and unfortunate side effect. Scientists placed inordinate conviction in the potential of this new science. As science historian Evelyn Fox Keller wrote: "During almost 50 years, we were convinced that, by discovering the molecular basis of genetic information, we will have uncovered the "secret of

© *The Auk*/American Ornithologists' Union

Once solicited to replace Ernst Hartert (see page 161) at the museum in Tring, Ernst Mayr (1904-2005) made his career at the American Museum of Natural History in New York. He had a strong influence on a number of ornithologists and contributed to greater scientific rigour in the observation of birds in the United States. Famous for his work on the biology of evolution and the concept of the species, Mayr played a considerable role in the development of the synthetic theory of evolution.

life". Molecular biology, which would dominate many research budgets to the detriment of traditional natural history, is only one piece of the puzzle in understanding of the "secret of life".

The development of
popular ornithology

Bird protection societies such as the **Royal Society for the Protection of Birds (RSPB)** in the United Kingdom (which has more than one million members), the **National Audubon Society** in the United States and the **Ligue Pour la Protection des Oiseaux (LPO**) in France are enduring symbols of the profound transformations that have occurred in the practice of ornithology during the 20th century.

At the end of the 19th century, the study of birds took place mainly through their breeding or collection and was accessible to only a small number of often very well-off people. At the beginning of the 20th century, the hunting and collecting of specimens began to give way to the observation of living birds: wildlife photography began to gain in popularity, as did ringing (or banding), which allowed the release of the birds after their capture and the future tracking of their behaviour and migration patterns.

In 1901, the loosely affiliated individual Audubon societies became closer and formed the National Committee of the Audubon Societies. Four years later they reorganized into what was called the National Association of Audubon Societies for the Protection of Wild Birds and Animals, and in 1940 the name was shortened to simply the National Audubon Society. The society began to work to give birds legal protection, with early successes including the Federal Migratory Bird Treaty of 1918 and the New York State Audubon Plumage Law of 1910. It also began to buy areas of land and designate them as wildlife sanctuaries, and it encouraged the government to do the same, which resulted in the development of the National Wildlife Refuge system. Later in the 20th century, the Audubon Society continued to successfully promote the welfare of birds and their habitats using these methods, and also through pushing for environmentally friendly legislation such as that to prevent drilling for oil in the Arctic National Wildlife Refuge and logging in the national forests.

There was a similar pattern of events in Britain, where the RSPB, which was formed to counter the trade in plumes for women's hats, bought areas of land and designated them as nature reserves to help populations of rare and threatened species such as the Pied Avocet and Marsh Harrier. This continued throughout the century and today the society has more than 200 reserves and 1,300 staff who work in a variety of ways, from managing reserves to dealing with wildlife crime and lobbying government over issues that affect wild birds.

In contrast to countries such as Britain, Germany and the USA, where ornithological societies organized themselves nationally from the 19th century, other countries such as France had no such structure and had to make up for lost time in the 20th century. In that country the Muséum National d'Histoire Naturelle in Paris played a pivotal role in the development of modern ornithology. **Jacques Paul Antoine Berlioz** (1891–1975), great great-nephew of Hector Berlioz, was captivated by natural history as a child and built up a collection of birds by the age of eight. He joined the museum in 1920 and became a professor and eventually took up the chair of zoology in 1949. **Jean Dorst** (1924–2001), no doubt influenced by his father who was a keen amateur butterfly collector, started collecting plants and animals from an early age. He began to frequent the museum from 1942. He succeeded Berlioz in 1964 and directed the museum from 1976 to 1985. A field naturalist, Dorst participated in a number of expeditions to Africa, the Andes and the Galapagos. With several biologists, including Sir Julian Huxley (see page 200), Sidney Dillon Ripley (1913–2001) and Victor Van Straelen (1889–1964), he founded the Darwin Foundation in order to establish a permanent research station in the Galapagos. He was the author of more than 600 publications, including *Avant que nature meure* (1965), which was translated into 17 languages and which contributed to a greater sensitivity to the threats facing the environment.

In 1909, Louis Denise (1863–1914), librarian at the Bibliothèque Nationale, and Auguste Ménégaux (1857–1937), assistant at the museum in Paris, founded the first French-language ornithological journal, the *Revue française d'ornithologie scientifique et practique*.

Interested in animals from a young age, Jean Théodore Delacour (1890–1985) went on to study medicine at Lille. In 1920, he launched *Revue d'histoire naturelle appliquée: l'Oiseau* for the Société d'Acclimatation, of which he was an active member. This publication was squarely focused on aviculture. The *Revue française d'ornithologie* and *L'Oiseau* merged in 1928. The new journal consisted of two parts, one scientific and the other devoted to aviculture. Another group of naturalists, which included Henri Heim de Balsac (1899–1973), launched the journal *Alauda* in reponse to this, seeking to be "the first dedicated ornithological journal" and to distance itself from aviculturalists. Each journal had its own society: *L'Oiseau et la Revue française d'ornithologie* was published by the Société Ornithologique de France (1930) and *Alauda* by the Société d'Études Ornithologiques (1933). These societies would merge in 1994.

In addition to these organizations there was the Ligue Pour la Protection des Oiseaux. This had its roots in the Société d'Acclimatation (which changed name to the Société Nationale de Protection de la Nature in 1960), which was established in 1854 and directed its activities towards the protection of the environment. Its division devoted to the protection of birds became an autonomous body in 1912: the Ligue Pour la Protection des Oiseaux. Its first nature reserve was on the Sept-Îles, and was created in 1913 to conserve the Atlantic Puffin (*Fratercula arctica*), which was being heavily hunted there at the time. A very active society, it organized the International Congress for the Protection of Nature in Paris (1923) and created the reserves of Camargue (1927), Néouvieille and Lauzanier (both 1935). Finally, it launched the journal *La Terre et la Vie* (1931), which aimed to raise awareness of conservation and natural history. It also participated in the founding of the International Union for Conservation of Nature (IUCN) in 1948.

In 1922 a number of national bird preservation societies such as the RSPB and LPO joined forces to form the International Council for Bird Preservation (ICBP), with the aim of protecting birds and their habitats. This global partnership, which was renamed BirdLife International in 1993, achieved significant success in protecting birds worldwide. By the end of the 20th century it had nearly 100 member organizations with more than 10 million supporters and 1 million hectares of land owned or managed for birds.

The end of the Second World War coincided with the emergence of an urban middle class in Western Europe and North America with leisure time at its disposal and keen to reconnect with nature. Birdwatching became a very popular recreational activity, and many local members' groups formed so that knowledge and experiences could be shared. The number of ornithological organizations grew rapidly to a much higher number than at any point previously in the history of the study of nature. The observation of birds was often linked to concern for the protection of species and habitats, and the activities of societies such as the LPO and RSPB were a good example of this.

This concern for the protection of birds would grow throughout the 20th century. Particularly from the 1960s onwards, the number of conventions, laws and national directives in Europe and around the world would multiply: in

The Ligue pour la Protection des Oiseaux (LPO), with its 40,000 members, is one of the most important French ornithological organizations. The LPO publishes the *L'Oiseau* journal (since 1985) and *Ornithos : revue d'ornithologie de terrain* (since 1994).

1973, the Convention on International Trade in Endangered Species of Wild Fauna and Flora (CITES); in 1979, the Bonn Convention on migratory species, the Berne Convention on the conservation of European wildlife and habitats, and the Birds Directive on wild birds; in 1992, the EU Habitats Directive.

In addition to the rise in bird protection societies, the popular interest in ornithology was also encouraged by an increase in the volume and quality of literature available. In addition to journals, the later part of the century saw the publication of popular magazines such as *Birdwatch* and *Bird Watching* in the UK and *Bird Watcher's Digest* in the USA, while there was a big increase in the number and quality of bird books available, especially field guides. Roger Tory Peterson (see page 192) published his first North American field guide in 1932 and, together with Guy Mountfort and P.A.D. Hollom, a European guide in a similar format in 1954 which remained in print for more than 50 years. The field guide format was refined and developed over the years and by the turn of the century saw such polished works as the *Collins Bird Guide on the Western Palearctic* by Lars Svensson, Peter J. Grant, Killian Mullarney and Dan Zetterstrom and *The North American Bird Guide* by David Sibley. Virtually every corner of the world was covered by a practical and illustrated field guide to its avifauna, while there was also a wealth of other books on all aspects of ornithology, from detailed monographs on species or families to bird-finding site guides, and from checklists of species and subspecies to handbooks, the most comprehensive example of the last being Lynx Edicions' mammoth 16-volume *Handbook of the Birds of the World*, the first part of which was published in 1992.

More so than ornithology itself, though, it was the bird that changed in status in the 20th century. It became a living symbol of the beauty of a disappearing idealized nature and the need to protect it against the ravages of urbanization, industrialization, agricultural modernization, pollution and climate change brought about by the rapidly multiplying human population.

Conclusion

Everyday and lifelong companions, birds have fascinated poets and captivated scientists. For a long time, the study of birds consisted of hunting and preserving skins. Eggs and nests were equally sought after as they did not require the mastery of taxidermy. Collections, the great passion of the 19th century, became highly coveted and attracted a great deal of attention.

Regardless of the period, scientific research must continually justify its utility. Botany provides a good example: the use of plants, be they for nutritional, industrial or medicinal reasons, justified study of new species. This was not the case with ornithology: for a long time, the primary justification for the study of new species would be their beauty and the pleasure of their contemplation. There were nonetheless numerous attempts to acclimatize exotic species. Apart from turkeys and guineafowls, most of these birds have had little economic impact.

The 20th century was a time of upheavals in ornithological practice: the gun was gradually abandoned, the gathering of eggs regulated. Having become unarmed observers of nature, ornithologists studied the behaviour of birds. The changes in practices gave rise to the immense popularity of ornithology which lends itself more to hobby than science. Its importance became, from then on, much larger.

The role and function of scientific ornithology

It is estimated that since 1600, 90 species and 60 subspecies of bird have become extinct. The causes are varied: the destruction of habitats, the introduction of predators and diseases (particularly on islands), the introduction of competing species, and overhunting (as in the cases of the Dodo, American Passenger Pigeon and Great Auk). Extinct species are often only known through specimens preserved in museums. It is difficult to come up with an accurate picture of the number of specimens preserved by the museums throughout the world, but it must be close to ten million. For comparative purposes, the estimated number of deaths resulting from the collision of birds against windows in the United States is in the range of a hundred million to a billion individuals each year. The sample taken by scientists thus only represents a very small impact on populations, lagging far behind that of others causes of mortality.

The British Museum now holds the world's largest avian collection with more than two million specimens. Consisting of skins, eggs, nests and skeletons, these collections cover 95% of all known species. Located in the town of Tring, 30 miles north-west of London, the collections are not open to the public but can be accessed on request by amateur and professional researchers. A portion of the complex is nonetheless open to the public, with 4,000 stuffed animals available for viewing.

Five institutions in the United States hold the next largest collections: the American Museum of Natural History in New York (more than a million specimens, including almost the entire Lord Rothschild collection), the National Museum of Natural History in Washington (more than 625,000 specimens), the Field Museum in Chicago (445,000 specimens), the Museum of Comparative Zoology at Harvard University in Cambridge (400,000 specimens) and the Museum of Zoology at the University of Michigan at Ann Arbor (roughly 200,000 specimens). Next are the collections of the Institute of Zoology at the Russian Academy of Science in Saint Petersburg (188,000 specimens), those of the four other American institutions and of the Royal Ontario Museum in Toronto, Canada. In Europe, noteworthy collections are held at the Royal Museum for Central Africa in Tervuren in Belgium (with 150,000 specimens) and the Museum für Naturkunde at the University of Humboldt in Berlin (with 150,000 specimens). The Muséum National d'Histoire Naturelle in Paris holds 130,000 skins, 30,000 mounted specimens and 6,000 skeletons.

The preservation of specimens in museums is not always guaranteed. As we have seen, some historic collections have been completely or largely lost like those of Sir William Jardine, Allan Hume, C.L. Brehm and many others. Natural catastrophes have taken their toll on some institutions: the museum of the Academy of Sciences in Chicago was completely destroyed in the great fire of 1871; the earthquake in San Francisco and the fire that followed destroyed the collection of the Academy of Sciences of California.

Many German and Japanese collections were lost during the Second World War, as were those of the museums of Colombo (Sri Lanka) and Calcutta (India). The events in Hungary in 1956 gave rise to the destruction of the museum in

Budapest which housed 70,000 bird skins. More recently troubles in the Caucasus brought about poor preservation and the eventual loss of the collection of the Academy of Sciences at Tbilisi in Georgia.

There were, unfortunately, some cases in which ill-intentioned people stole specimens or altered collections. One example was that of an amateur who, from 1975 to 1979, stole more than 30,000 eggs from the British Museum, and who, to cover up his trail, inverted the labels used for their identification. Roughly 10,000 eggs were retrieved following his arrest. His actions permanently diminished the largest egg collection in the world.

Today we continue to enrich these collections with passive acquisitions (birds that died from natural causes, killed by moving vehicles, collisions with windows or caught by cats). It was in this way that the museum in Edinburgh received its first specimens of the Red-necked Stint (*Calidris ruficollis*) and the Thick-billed Warbler (*Acrocephalus aedon*). Similarly, specimens of exotic species are obtained from breeders or zoos, allowing for the completion of collections. Sadly, some specimens have been acquired following oil slicks

The preservation and study of established collections, like the acquisition of new specimens, play important roles in ornithology. Firstly, they allow for the establishment of reference collections used in the identification of species and systematics. The study of old specimens can serve many applications, among them the study of the effect of DDT on the thickness of eggshells. Today, collections are used for the historical references that they provide, including the monitoring of population changes over long periods.

Ornithology, still a topical subject of study?

The study of birds led to many advancements in several disciplines of biology including ethology and evolution. Without a doubt, it has been in the field of ecology and its practical applications (the conservation of the environment) that the contributions of ornithology have been the greatest.

Birds are one of the few zoological groups of which we know, it seems, almost all species. What's more, the habits and ecological requirements of most European and North American species are well known. This was the work of numerous ornithologists, who, over the course of centuries carried out relentless work and created efficient organizations. To this we can add the numerous amateur birders who have

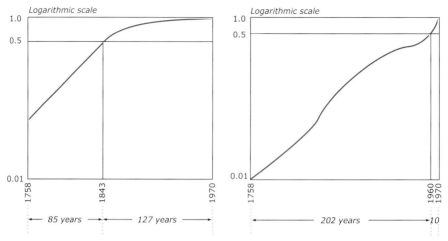

These two curves demonstrate the progression of the number of species discovered in the last two and a half centuries. **To the left, birds**. The curve shows that in 1843 we knew nearly half the species we know today. The rate of discovery slowed progressively to the point where it barely changes. **To the right, arachnids and molluscs**. The curve shows that half of known species have only been discovered very recently (1960). Discoveries continue to this day at a rapid rate.

been active proponents of the environmental movements that emerged in the 1970s. They have helped to change minds and legislation.

In Europe and in France in particular, there has been a general feeling that professional scientific research in the traditional fields of natural history suffers from multiple ills, including small budgets and lack of recognition, particularly from political leaders. The situation is especially bad in disciplines such as systematics, where the numbers of professionals retiring are not being replaced. The teaching of natural history is regressing (where it has not completely disappeared) and certain botanical gardens connected to pharmacy departments are under threat of closure. The unfortunate state of these disciplines, often considered out of fashion, is occurring, paradoxically, during the period in which the threats to the natural world and to biodiversity have never been greater.

Nonetheless, ornithology seems to have weathered this crisis without suffering too much damage or weakening of its popularity. It will probably take on a new role in the monitoring of the progress of climate change, as changes in the geographic distribution of bird populations serve as an excellent indicator of this.

BIBLIOGRAPHY

Individual articles and biographical works are not cited.

- ● History of ornithology

Mark V. Barrow (1998). A *Passion for Birds. American Ornithology after Audubon*, Princeton University Press (Princeton, New Jersey): 336 pp.

John Thomas Battalio (1994). *The Professionalization of a Science: Rhetoric and the Evolution of Expertise in Ornithological Discourse*, 1995, 55, 1927-A. Numbers Univ. Microfilms order no. 94-32643. Dissertation at Northwestern University, Dissertation Abstracts International collection: XI + 179 pp.

Bo Beolens and Michael Watkins (2003). *Whose Bird? Common Bird Names and the People They Commemorate.* Christopher Helm (London): 400 pp.

Maurice Boubier (1925). *L'Évolution de l'ornithologie.* Librairie Félix Alcan (Paris), Nouvelle collection scientifique: II + 308 pp.

Robin W. Doughty (1975). *Feather Fashions and Bird Preservation*, California University Press (Berkeley): XI + 184 pp.

Paul Lawrence Farber (1996). *Discovering Birds: the Emergence of Ornithology as a Scientific Discipline*, 1760-1850, Johns Hopkins University Press (Baltimore), Studies in the history of modern science collection: XXIII + 191 pp.

Edward S. Gruson (1972). *Words for Birds. A Lexicon of North American Birds with Biographical Notes.* Quadrangle Books (New York): XIV + 305 pp.

Joseph Kastner (1986). *A World of Watchers. An Informal History of the American Passion for Birds, Sierra Club Books* (San Francisco): X + 241 pp.

Barbara Mearns & Richard Mearns (1988). *Biographies for Birdwatchers. The Lives of Those Commemorated in Western Palearctic Bird Names.* Academic Press (London): XX + 490 pp.

Barbara Mearns & Richard Mearns (1992). *Audubon to Xantus. The Lives of Those Commemorated in North American Bird Names,* Academic Press (London): XIX + 588 pp.

Barbara Mearns & Richard Mearns (1998). *The Bird Collectors.* Academic Press (London): XVII + 472 pp.

David Snow (1992). *Birds, Discovery and Conservation: 100 Years of the 'Bulletin of the British Ornithlogists' Club'*, Helm Information (Mountfield): IX + 198 pp.

Erwin Stresemann (1975). *Ornithology from Aristotle to the Present.* Cambridge University Press (Cambridge): XII + 432 pp.

Michael Walters (2003). *A Concise History of Ornithology.* Christopher Helm (London): 255 pp.

- ● History of illustration

Stanley Peter Dance (1990). *Les Oiseaux*, Celiv: 128 pp.

John C. Devlin and Grace Naismith (1977). *The World of Roger Tory Peterson. An Authorized Biography,* Optimum Publishing Company Limited (Ottawa): XXI + 266 pp.

Jonathan Elphick (2004). *Les Oiseaux*, Mengès (Paris): 335 pp.

Brian John Ford (1993). *Images of Science: a History of Scientific Illustration,* Oxford University Press: VIII + 208 pp.

Robert Guinot (2002). *Jacques Barraband: le peintre des oiseaux de Napoléon Ier*, Guénégaud (Paris): 191 pp.

Susan Hyman (1980). *Edward Lear's Birds*, Trefoil (London): 96 pp.

Christine E. Jackson (1985). *Bird Etchings. The Illustrators and Their Books*, 1655-1855, Cornell University Press (Ithaca): 292 pp.

Joseph Kastner and Miriam T. Gross (1988). *The Bird Illustrated, 1550-1900. From the Collections of the New York Public Library*, Abrams H.N. (New-York): 128 pp.

Joseph Kastner and Miriam T. Gross (1991). *The Animals Illustrated, 1550-1900. From the Collections of the New York Public Library*, Abrams H.N. (New-York): 128 pp.

David M. Knight (1977). *Zoological Illustration: an Essay Towards a History of Printed Zoological Pictures*, Dawson (Folkstone, Kent) and Archon books: XII + 204 pp.

Maureen Lambourne (2001). *The Art of Bird Illustration. A Visual Tribute to the Lives and Achievements of the Classic Bird Illustrations*, Eagle Editions Ltd (Royston): 192 pp.

A.M. Lysaght (1985). *The Book of Birds: Five Centuries of Bird Illustration*, Bookthrift Co: 208 pp.

Robert McCracken Peck (1982). *A Celebration of Birds. The Life and Art of Louis Agassiz Fuertes*, Walker and Company (New York): XIII + 178 pp.

Francis Roux (1993). *Les Oiseaux exotiques de John Gould*, Bibliothèque de l'image (Paris): 96 pp.

Francis Roux (1992). *Le Livre des Oiseaux. Audubon*, Bibliothèque de l'image (Paris): 96 pp.

Francis Roux (2000). *Les Oiseaux d'Europe de John Gould*, Bibliothèque de l'image (Paris): 239 pp.

- History of natural history

Alan Axelrod and Charles Phillips (1993). *The Environmentalists: a Biographical Dictionary from the 17th Century to the Present*, Facts on File (New York): XIV + 258 pp.

Paul Delaunay (1997). *La Zoologie au XVIᵉ siècle*, Hermann (Paris): XI + 338 pp.

Paul Lawrence Farber (2000), *Finding Order in Nature. The Naturalist Tradition from Linnaeus to E.O. Wilson*, Johns Hopkins University Press (Baltimore): 160 pp.

Joseph Kastner (1978). *A World of Naturalists*, John Murray (London): XIV + 351 pp.

Joseph Kastner (1969). *A Species of Eternity*, Dutton (New York): XIV + 350 pp.

David M. Knight (1981). *Ordering the World. A History of Classifying Man*, Burnett Books (London): 215 pp.

Arthur Laurent (2006). *Pionniers de la photographie animalière*, Pôles d'images (Barbizon): 175 pp.

Patrick Matagne (2002). *Comprendre l'écologie et son histoire*, Delachaux and Niestlé (Paris), La Bibliothèque du naturaliste collection: 208 pp.

Michael Andrew Osborne (1994). *Nature, the Exotic, and the Science of French Colonialism*, Indiana University (Bloomington Press): XVI + 216 pp.

Keir B. Sterling, Richard P. Harmond, George A. Cevasco and Lorne F. Hammond (1997). *Biographical Dictionary of American and Canadian Naturalists and Environmentalists*, Greenwood Press (Westport): XIX + 937 pp.

Mary P. Winsor (1976). *Starfish, Jellyfish, and the Order of Life. Issues in Nineteenth-Century Science*, Yale University Press (New Haven, Connecticut): X + 228 pp.

TIMELINE

Science

Biodiversity

Voyages and Discoveries

Technical Advancements

Historical Points of Reference

Institutions and Societies

| 1000 BC | *Antiquity*

500 BC.

Study of natural history
begins with Aristotle

*c.*340 BC: **Aristotle** (384–322 BC) describes 140 species of
birds in his *History of Animals*.

0

*c.*70: **Pliny the Elder** (23–79) wrote his
Natural History

Pliny the Elder

2nd century: ***Physiologos***, Greek bestiary by an
unknown author.

370: ***Physiologus***, Latin adaptation of *Physiologos*.

Physiologus

500

The Middle Ages

*c.*630: **Isidore of Seville** (*c.*560–636)
composed his *Etymologiae*.

Isidore of Seville
(right)

From 800 to 1100, the
translation of Aristotle's
works into Arabic

*c.*850: **Al–Jahiz** (*c.*776–868 or 869) wrote his works on
animals.

1000

1000 —

11th and 12th centuries: **Avicenna** (980–1037) and
Averroes (1126–1198) translated and commented on
Aristotle's texts.

Avicenna

1100 —

1200 —

*c.*1230: Latin translation of Aristotle's work by **Michael Scot**
(*c.*1175–*c.*1232).

*c.*1240: **Frederick II of Hohenstaufen** (1194–1250)
composed his *Arte venandi cum avibus*, which contains many
precise ornithological observations.

Frederick II

*c.*1270: **Albertus Magnus** (*c.*1200–1280) translated
Aristotle's work into Latin as *De animalibus*. **Thomas
Cantimpré** (1201–1272) described 144 birds in his *De
naturis rerum*. **Vincent de Beauvais** (*c.*1199–*c.*1265)
described farmyard birds in his *Miroir de la Nature*.

Albertus Magnus

221

TIMELINE

1300 —

The Renaissance

*c.*1400: extinction of Haast's Eagle (*Harpagornis moorei*), a
giant species from New Zealand.

1400 —

*c.*1400: the invention
of printing

1475: *Buch der Natur*, first book printed depicting birds.

1485: *Gart der Gesundheit* by **Johann Wonnecke von Caub**,
first illustrated natural history book. One of its parts, *De
avibus*, dealt with birds, bats and insects.

1492: discovery of
America; end of the
Spanish Reconquista

Buch der Natur

1500 —

1500 —

Growth in the number
of aviaries, often with
exotic species

1530: the publication of
Otto Brunsfels's herbarium
showed progress in the
realism of illustrations

1544: *Avium praecipuarum* by **William Turner** (*c*.1510–
1568), first book entirely devoted to birds.

1544: *Dialogus de avibus et earum nominibus graecis, latinis et
germanicis* by **Longolius** (1507–1543).

1543: creation of the first
botanical garden at Pisa

222

1550 —

1555: *Histoire de la nature des oyseaux* by **Pierre Belon**
(*c*.1517–1564) presented a number of original observations
and gave the first known formula for the preservation of bird
skins.

1555: Belon and Gessner
described 222 species

1555: *Historiae animalium* by **Conrad Gessner** (1516–
1565); the third volume is devoted to birds.

*Avium
praecipuarum*

Pierre Belon   Conrad Gessner

1572–1574: **Francisco Hernández** (1514–1578) conducted
the first scientific expedition to Central America.

1575: *De avium sceletis et praecipius musculis* by **Volcher
Coiter** (1534–1576). He studied the anatomy of birds and
would later take an interest in the embryonic development
of eggs.

Volcher Coiter

1593: Creation of
Montpellier botanic garden

1599–1603: *Ornithologiae hoc est de avibus historiae*, book XII
by **Ulisse Aldrovandi** (1522–1605).

Ulisse Aldrovandi

1600 —

The Development of Anatomy

1600 —

1600: Girolamo Fabrizio or **Fabricius** (1537–1619) was interested in the embryonic development of birds.

1603: *Theriotropheum Silesiae* by **Caspar Schwenckfeld** (1563–1609); first work on regional fauna.

Emperor Rudolph II (1552–1612) assembled a vast cabinet of curiosities and a menagerie

1605: *Exoticorum libri decem* by **Charles de l'Écluse** (or **Carolus Clusius**) (1525–1609), first work devoted solely to exotic fauna.

Charles de l'Écluse

1600 and 1602: formation of the British and Dutch East India Companies

1622: *Uccelliera, overo discorso della natura e proprietà di diversi uccelli e in particolare di que' cha cantano* by **Giovanni Pietro Olina** (1587–1645), written for owners of aviaries.

Olina's *Uccelliera*

1625 —

1628: the discovery of blood circulation by **William Harvey** (1578–1657).

223

TIMELINE

1635: creation of the Jardin du Roi in Paris

1645: *Zootomia Democritaea* by **Marco Aurelio Severino** (1580–1656).

1648: *Historia naturalis Brasiliae* by **Georg Markgraf** (1610–1648) and **Willem Piso** (1611–1675), the first study of the fauna and flora of South America; 133 species of birds are described.

Historia naturalis Brasiliae

1650 —

1650 ——

1650: the microscope was first used as a scientific instrument

1651: *Exercitationes de generatione animalium* by **William Harvey** (1578–1657) in which he studied the embryonic development of chicken eggs as well as the anatomy of these birds.

1655: *Museum Wormianum* by **Ole Worm** (1588–1654), catalogue of his cabinet of curiosities.

1657: *Historiae naturalis*, volume IV, *De avibus*, by **John Jonston** (1603–1675).

1664: **Robert Hooke** (1635–1703) published *Micrographia* in which he described feather barbules.

1666: *Pinax rerum naturalium Britannicarum,* by **Christopher Merrett** (1614–1695), the first study of British fauna.

1666: creation of French Académie des Sciences

1667–1671: **Jean–Baptiste Du Tertre** (1610–1687) described the birds he observed in *Histoire générale des Antilles*.

1668: *Onomasticon Zoicon* by **Walter Charleton** (1619–1707).

1671: **Friderich Martens** published the account of his voyage to Spitsbergen and was the first to describe species of birds from the extreme northern regions of the Atlantic.

1672: **Marcello Malpighi** (1628–1694) described embryonic development in chicken eggs.

1675 ——

Ray and Willughby

John Ray and Francis Willughby

1676: *Ornithologia* by **Francis Willughby** (1635–1672), edited by **John Ray** (1627–1705). Described roughly 500 species. For the first time, classification was based on anatomy.

17th and 18th century: extinction of a number of species from the Mascarenes, including the Rodrigues Solitaire (*Pezophaps solitaria*), Rodrigues Grey Pigeon (*Alectroenas rodericana*), Rodrigues Night-heron (*Nycticorax megacephalus*), Reunion Shelduck (*Alopochen kervazoi*), Mauritian Shelduck (*Alopochen mauritianus*), Red Rail (*Aphanapteryx bonasia*), Rodrigues Rail (*Aphanapteryx leguati*), Mascarene Coot (*Fulica newtoni*), Broad-billed Parrot (*Lophopsittacus mauritianus*), Reunion Owl (*Mascarenotus grucheti*), and Mauritius Owl (*Mascarenotus sauzieri*).

1700 ——

1700 —

*c.*1700: **Georg Eberhard Rumpf** or **Rumphius** (1627–1702) made a number of ornithological observations in Indonesia.

1702: **Johann Ferdinand Adam von Pernau** (1660–1731) published his first book on the behaviour of birds: *Unterricht was mit dem lieblichen Geschöpff, denen Vögeln, auch ausser dem Fang Nur durch die Ergründung Deren Eigenschafften und Zahmmachung oder ander Abrichtung Man sich vo Lust und Zeitvertreib machen könne: gestellt Durch den Hoch- und Wohlgebohrnen* (Instructions on the pleasures that can be obtained from these charming creatures, birds, not to mention their capture as well as a complete study of their habits and efforts to domesticate and to instruct them).

1713: *Synopsis methodica avium et piscium* by **John Ray** (1627–1705), published after his death thanks to **William Derham** (1657–1735).

1725 —

1726: *Oud en Nieuw Oost Indiën* by **François Valentijn** (1666–1727), description of fauna based on the descriptions of Georg Eberhard Rumpf.

1731–1748: *Natural History of Carolina, Florida and the Bahama Islands* by **Mark Catesby** (1683–1749), the first descriptions of the birds of North America.

Catesby's work

1731–1738: *A Natural History of Birds* by **Eleazar Albin** (?–1741?) illustrated with 306 engravings.

1734–1765: **Albertus Seba** (1665–1736) published the catalogue of his cabinet of curiosities, *Locupletissimi rerum naturalium thesauri accurata descriptio.*

Albertus Seba

1737: *Delle uova e dei nidi degli uccelli libro primo del Conte Giuseppe Zinanni. Aggiunte in fine alcune osservazioni, con una dissertazione sopra varie spezie di Cavalette* by Count **Giuseppe Zinanni** (1692–1753), the first book entirely devoted to birds' eggs.

Eleazar Albin

1742–1743: *Petino-Theologie, oder Versuch die Menschen durch naehere Betrachtung der Voegel zur Bewunderung, Liebe...* by **Johann Heinrich Zorn** (1698–1748) who continued work and observations started by Pernau.

1743–1751: *A Natural History of Uncommon Birds, and of Some Other Rare and Undescribed Animals* by **George Edwards** (1694–1773).

1745: *Ornithologiae specimen novum* by **Pierre Barrère** (*c.*1690–1755).

George Edwards

1750 —

1750

1750: *Historiae avium prodromus cum praefatione de ordine animalium in genere* by **Jacob Theodor Klein** (1685–1759).

1752: *Avium genera* by **Paul Heinrich Gerhard Möhring** (1710–1792).

1753: **Sir Hans Sloane** (1660–1753) bequeathed his collections to the Crown, which would become the beginnings of the British Museum.

1755

Sir Hans Sloane

1757: Linnaeus lists five bird orders with 564 species

The Linnaean Classification

1760: Brisson described 26 orders, 150 genera and 1,500 species and subspecies.

1757: *Systema naturae* by **Carl Linnaeus** (1707–1778), tenth edition of his major work (the starting point for zoological nomenclature).

Systema Naturae

1760

1760: *Ornithologie* by **Mathurin Jacques Brisson** (1723–1806).

1761: *British Zoology* by **Thomas Pennant** (1726–1798).

1764: *Ornithologia borealis* by **Morten Thrane Brünnich** (1737–1827).

1764: **Charles Georges Leroy** (1723–1789) began to publish his papers on the intelligence and sensitivity of animals, considered one of the first ethological works.

1765

Mathurin Jacques   Thomas
Brisson   Pennant

Morten Thrane
Brünnich

226

TIMELINE

1766-1769: Bougainville's voyage

1768-1771: first circumnavigation of the globe by James Cook

1769: *Outlines of the Natural History of Great Britain* by **John Berkenhout** (1730–1791).

1769: *Indian Zoology* by **Thomas Pennant** (1726–1798).

1770–1783: *Histoire naturelle des oiseaux* by **Georges–Louis Leclerc, Count of Buffon** (1707–1788) in nine volumes.

1770

1771: *Ornithologia britannica* by **Marmaduke Tunstall** (1743–1790) in which he described British birds following the Linnaean method.

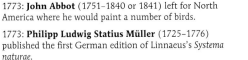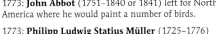

Buffon

1773: **John Abbot** (1751–1840 or 1841) left for North America where he would paint a number of birds.

1773: **Philipp Ludwig Statius Müller** (1725–1776) published the first German edition of Linnaeus's *Systema naturae*.

John Abbot

1775

1775 —

1776: *Gli Ucelli di Sardegna* de **Francesco Cetti**
(1726-1778).

1779: *Elementa ornithologica* by **Jacob Christian Schäffer**
1780 — (1718-1790).

Jacob Christian
Schäffer

1781: the beginning of **Félix de Azara's** (1746–1811) stay
in South America, where he made numerous ornithological
discoveries.

1782: *Voyage aux Indes orientales et à la Chine* by **Pierre
Sonnerat** (1748–1814).

1783: article on ornithology in *Encyclopédie méthodique* by
P.J.C. Mauduyt de la Varenne (ca. 1732–1792).

Pierre Sonnerat

1783: **Pieter Boddaert** (1730–1796) ascribed Linnaean
scientific names to the birds described by Buffon.

1784–1785: *Arctic Zoology* by **Thomas Pennant** (1726–
1785 — 1798).

1785-1788: voyage of Jean-
François de La Pérouse

1788–1793: **Johann Friedrich Gmelin** (1748–1804)
published the 13th edition of *Systema naturae* by Linnaeus.

Johann Friedrich
Gmelin

1788–1792: **Abbot Pierre Joseph Bonnaterre** (1747–
1804) was the first French ornithologist to adopt the
Linnaean system.

1788: facsimile of Frederick II's Arte venandi cum avibus
by **Blasius Merrem** (1761–1824) and **Johann Gottlob
Schneider** (1750–1822).

1789: French Revolution

1789: *Natural History and Antiquities of Selborne* by **Gilbert
White** (1720–1793).

1790 —

1790: *Index ornithologicus sive Systema ornithologiae* by **John
Latham** (1740–1837), in which he described 2,951 species
and subspecies of birds.

John Latham

1793: creation of the
French Muséum National
d'Histoire Naturelle

1791: *Travels through North and South Carolina* by **William
Bartram** (1739–1823).

William Bartram

1795 —

1795–1803: *Naturgeschichte der Vögel Deutschlands* by
Johann Andreas Naumann (1744–1826) and **Johann
Friedrich Naumann** (1780–1857).

1796–1808: publication of several works by **François
Levaillant** (1753–1826): *Histoire naturelle des oiseaux
d'Afrique* (1796–1808), *Histoire naturelle d'une partie d'oiseaux
nouveaux et rares de l'Amérique et des Indes* (1801), *Histoire
naturelle des oiseaux de paradis* (1801–1806), *Histoire naturelle
des perroquets* (1801–1805), *Histoire naturelle des cotingas et
des todiers* (1804), *and Histoire naturelle des calaos* (1804).

Johann Andreas Naumann and
Johann Friedrich Naumann

1799-1804: Humboldt
and Bonpland's voyage
to South America

1797: *British Birds* by **Thomas Bewick** (1753–1828).

François
Levaillant

Thomas
Bewick

1800 —

1800

1802: *Ornithogical Dictionary; or Alphabetical Synopsis of British Birds* by **George Montagu** (1753–1815).

1802: *Ornithologisches Taschenbuch von und für Deutschland oder Kurze Beschreibung aller Vögel Deutschlands für Liebhaber dieses Theils der Naturgeschichte* by **Johann Matthäus Bechstein** (1757–1822).

1803–1804: **Jean–Baptiste Bécoeur's** (1718–1777) arsenic-based formula for the preservation of bird skins was popularized in *Nouveau Dictionnaire d'histoire naturelle de Louis Dufresne* (1752–1832)

George Montagu

1805

1808–1814: *American Ornithology* by **Alexander Wilson** (1766–1813), first work on the regional fauna of North America.

Alexander Wilson

TIMELINE

1810

1810: founding of the Berlin Museum

1811: *Catalogue des oiseaux du Piémont* by **Franco Andrea Bonelli** (1784–1830).

1811: *Zoographia Rosso-Asiatica*, a posthumous publication by **Peter Simon Pallas** (1741–1811) and his primary work.

1812: **Johann Karl Wilhelm Illiger** (1775–1813) described 3,779 species and subspecies of birds.

Peter Simon Pallas   Franco Andrea Bonelli

1813: **Blasius Merrem** (1761–1824) presented his ideas on the classification of birds to the Academy of Sciences in Berlin.

Blasius Merrem

1815

1816: *Analyse d'une nouvelle ornithologie élémentaire* by **Louis Jean Pierre Vieillot** (1748–1831).

1815: end of the Napoleonic Wars

1817–1836: **Joseph Paul Gaimard** (1796–1858) and **Jean René Constant Quoy** (1790–1869) participated in several scientific missions around the world.

1820

1820-1840

1820

1820: *Manuel d'ornithologie, ou Tableau systématique des oiseaux qui se trouvent en Europe* by **Coenraad Jacob Temminck** (1778–1858), first edition.

Coenraad Jacob Temminck

1820: **Christian Ludwig Nitzsch** (1782–1837) published his work on the nasal glands of birds.

From 1820 to 1850: the Natuurkundige Commissie sent naturalists to the Dutch colonies in Asia

1820–1822: **Christian Ludwig Brehm** (1787–1864) published *Beiträge zur Vögelkunde*, in which he described 104 species of German birds.

1821–1825: the exploration of northeast Africa by **Friedrich Wilhelm Hemprich** (1796–1825) and **Christian Gottfried Ehrenberg** (1795–1876).

Christian Ludwig Brehm

1822: **William Bullock** (1773–1849) was the first naturalist to study the birds of Mexico.

1825

1825: *Ornithologie provençale* by **Polydore Roux** (1792–1833).

1825–1833: *American Ornithology* by **Charles Lucien Bonaparte** (1803–1857).

Charles Lucien Bonaparte   John James Audubon

1825–1843: *Illustrations of Ornithology* by **Sir William Jardine** (1800–1874).

1827–1831: *Ornitologia toscana* by **Paolo Savi** (1798–1871).

1827–1838: *Birds of America* by **John James Audubon** (1785–1851).

1828: *Manuel d'ornithologie* by **René Primevère Lesson** (1794–1849), followed by *Traité d'ornithologie* (1831).

Paolo Savi   George Ord

1828: **George Ord** (1781–1856) finished the publication of *American Ornithology* by **Alexander Wilson** (1766–1813).

1830

229

TIMELINE

1831: Charles Darwin's around the world voyage

1832: *Illustrations of the Family of the Psittacidae* by **Edward Lear** (1812–1888), his first work.

René Primevère Lesson

1832: *Manual of the Ornithology of the United States and Canada* by **Thomas Nuttall** (1786–1859).

1832–1837: *The Birds of Europe* by **John Gould** (1804–1881).

1833: *Schlesiens Wirbelthier-Fauna* by **Constantin Wilhelm Lambert Gloger** (1803–1863).

Thomas Nuttall

1835

1836: creation of the Australian Museum

1836: *A History of the Rarer British Birds* by **Thomas Campbell Eyton** (1809–1880).

1838-1842: discovery of Antarctica

1837: **William Swainson** (1789–1855) presented a classification of birds based on the quinary system.

1839: the first daguerreotype

William Swainson

1838: **Sir Richard Owen** (1804–1892) proved the existence of the moa.

1840

1840 —

1840–1848: *The Birds of Australia* by **John Gould** (1804–1881) following his voyage to Australia.

John Gould

1841: **George Robert Gray** (1808–1872) described 6,000 bird species and subspecies.

1842: **John Cassin** (1813–1869) joined the Academy of Natural Sciences of Philadelphia, for which he would build one of the largest ornithological collections.

1843: *The History of British Birds* by **William Yarrell** (1784–1856) became a reference work for British ornithologists.

John Cassin

1843: publication of the first version of the **Code of Zoological Nomenclature**, under the direction of **Hugh Edwin Strickland** (1811–1853).

1844: *Catalogue of Nipalese birds* by **Brian Houghton Hodgson** (1800–1894).

1844–1848: **George Robert Gray** (1808–1872) began to publish *List of the Genera of Birds*, in which he eventually listed all the genera of birds known at the time.

Brian Houghton Hodgson

Hugh Edwin Strickland

1845: founding of the journal **Rhea** by Ludwig Thienemann (1793–1858).

1844: extinction of the Great Auk (*Pinguinus impennis*)

230

1845 —

1846: *Gleanings from the Menagerie at Knowsley Hall* by **Edward Lear** (1812–1888).

Edward Lear

1849: **Edward Blyth** (1810–1873) published a catalogue on the birds of India.

1850 —

Edward Blyth

The observation of birds and the organization of ornithologists

1850 —

1850: creation of the **Deutsche Ornithologen– Gesellschaft (DO–G.**, or German Society of Ornithologists). The society had 107 members in 1853 and 230 in 1858.

1853: *Ornithologie de la Savoie, ou Histoire des oiseaux: qui vivent en Savoie à l'état sauvage soit constamment, soit passagèrement* by **Jean–Baptiste Bailly** (1822–1880).

1853: **Jean Louis Cabanis** (1816–1906) founded the *Journal für Ornithologie* which would become the official journal of the Deutsche Ornithologen–Gesellschaft the following year. It is the oldest ornithological publication in the world that remains in publication.

1854–1858: *De vogels van Nederland* by **Hermann Schlegel** (1804–1884).

1854: creation of French Société d'Acclimatation

1854–1862: **Alfred Russel Wallace** (1823–1913) travelled to the Malaysian Archipelago where he assembled a rich natural history collection.

Hermann Schlegel

1855 —

1855: **Robert Swinhoe** (1836–1877) began to study the birds of the China Sea.

TIMELINE

Robert Swinhoe

1857: **Philip Lutley Sclater** (1829–1913) published the first modern works of avian biogeography.

1858: founding of the **British Ornithologists' Union (BOU)** by Alfred Newton (1829–1907), Reverend Henry Baker Tristram (1822–1906) and ten other ornithologists.

BOU

1859: publication of *On the Origin of Species by Means of Natural Selection* by Charles Darwin

1859: founding of *The Ibis*, journal of British Ornithologists' Union.

1859: *Essais étymologiques sur l'ornithologie de Maine-et-Loire: ou les moeurs des oiseaux expliquées par leurs noms* by **Michel Honoré Vincelot** (1815–1877).

1860 —

1860 —

1860: the **Academy of Natural Sciences of Philadelphia** had built the largest ornithological collection in the world with 29,000 specimens.

1861: discovery of the first fossil Archaeopteryx.

1862: *The Birds of India* by **Thomas Claverhill Jerdon** (1811–1872).

Thomas Claverhill Jerdon

1861-1865: American Civil War

1864: the first issue of *Zoological Record* (which itemized all the scientific publications published in the year) cited only 120 ornithological publications.

232

1865 —

1865-1870: Gregor Mendel's experiments

1866–1872: **Père Armand David** (1826–1900) explored China.

1867: creation of the **Deutsche Ornithologische Gesellschaft** (DOG or German Ornithological Society) by Jean Louis Cabanis who disagreed with the members of the DO-G, created in 1850.

Père Armand David

1869: the **Sea Birds Preservation Act** protected 35 species of marine birds, first law for the protection of birds in Great Britain.

1869–1874: *Ornithologie Nord-Ost Afrikas* by **Theodor von Heuglin** (1824–1876).

1869–1872: **George Robert Gray** (1808–1872) published *Handlist of the Genera and Species of Birds*, in which he listed all the names of animal species.

George Robert Gray

1870 —

1870

1871: **Friedrich Hermann Otto Finsch** (1839–1917) began his voyages.

1871: **George Robert Gray** (1808–1872) described 11,162 species and subspecies of birds.

1872: the British Museum had acquired 30,000 bird skins.

1872: **Richard Bowdler Sharpe** (1847–1909) took on the direction of the department of ornithology at the British Museum.

1872: *Key to North American Birds* by **Elliott Coues** (1842–1899).

1873: founding of the **Nuttall Ornithological Club** whose aim was to facilitate the publication of scientific ornithological works. Its publication, *Bulletin of the Nuttall Ornithological Club*, began in 1876.

1873–1876: *Ornitologia italiana*, posthumous publication by **Paolo Savi** (1798–1871).

1873: **Anton Reichenow** (1847–1941) published his first work on the birds of Africa.

1874: **Count Adelaro Tommaso Paleotti Salvadori** (1835–1923) began to publish his works on the birds of Asia.

Friedrich Hermann Otto Finsch

Elliott Coues

1875

1875–1884: *History of North American Birds* by **Spencer Fullerton Baird** (1823–1887), **Thomas Mayo Brewer** (1814–1880) and **Robert Ridgway** (1850–1929).

Spencer Fullerton Baird

233

TIMELINE

1877: *Histoire naturelle des oiseaux-mouches, ou colibris constituant la famille des Trochilidés* by **Étienne Mulsant** (1797–1880) and **Édouard Verreaux** (1810–1868).

1878: extinction of the Labrador Duck (*Camptorhynchus labradorius*) in North America.

1880

The Protection of Birds

1880 —

1880: *Ornithologie de la Sarthe* (1880) by **Ambroise Gentil** (1842–1929).

1882: *Manual of the Birds of New Zealand* by **Sir Walter Lawry Buller** (1838–1906).

Sir Walter Lawry Buller

1883: *The Nests and Eggs of Indian Birds* by **Allan Octavian Hume** (1829–1912).

1883: the British Ornithologists' Union published *A List of British Birds*.

1883: foundation of the **American Ornithologists' Union** in September 1883 by Elliott Coues (1842–1899), Joel Asaph Allen (1838–1921) and William Brewster (1851–1919).

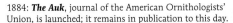

A List of   Allan Octavian
British Birds   Hume

1883: *A History of British Birds* by **Henry Seebohm** (1832–1895).

1884: **The Auk**, journal of the American Ornithologists' Union, is launched; it remains in publication to this day.

1884: the first **International Ornithology Congress** in Vienna, one the main themes was the protection of birds.

1885 —

1885: launch of the journal **Ornis**, the bulletin of the International Ornithological Committee.

1885: the first vaccination against rabies

234

TIMELINE

1887: **Daniel Giraud Elliot** (1835–1915) was influential in the creation of the American Museum of Natural History in New York.

Daniel Giraud Elliot

1889: founding of the **Royal Society for the Protection of Birds (RSPB)** in Great Britain, which sought to fight the trade of Great Crested Grebe feathers.

1889–1907: *Catalogue des oiseaux de la Suisse* by **Victor Fatio** (1838–1906) and **Theophile Rudolf Studer** (1845–1922).

1890 —

1890 —

1892: **Lord Lionel Walter Rothschild** (1868–1937) founded his museum at Tring, near London, **Ernst Hartert** (1859–1933) became its bird curator.

1893: **Hans Friedrich Gadow** (1855–1928) presented his classification of birds.

1893: beginning of the publication of *Dictionary of Birds* by **Alfred Newton** (1829–1907).

1894: the **Deutsche Ornithologen–Gesellschaft** bought back *Journal für Ornithologie*.

1894: *Lost British Birds* by **William Henry Hudson** (1841–1922).

Lionel Walter Rothschild   Ernst Hartert

Lost British Birds by W.H. Hudson

235

1895 —

1896: *Habit and Instinct* by **Conway Lloyd Morgan** (1852–1936) was one of ethology's founding works.

1896: **Louis Agassiz Fuertes** (1874–1927) published his first illustrations.

1898: *Wild Life at Home: How to Study and Photograph It* by **Richard Kearton** (1862–1928) and **Cherry Kearton** (1871–1940).

1899: launch of the American journal ***The Condor***, which is still in print today.

*c.*1900: **Hans Christian Cornelius Mortensen** (1856–1921) and **Johannes Thienemann** (1863–1938) tested a ringing (or banding) technique using aluminium rings.

1900 —

1900

1900: extinction of Brace's Emerald (*Chlorostilbon bracei*) and Gould's Emerald (*Chlorostilbon elegans*) in Jamaica and the Bahamas.

1900: *Bird Studies with a Camera: with Introductory Chapters on the Outfit and Methods of the Bird Photographer* by **Frank Michler Chapman** (1864–1945).

1901: **Johannes Thienemann** (1863–1938) settled in Eastern Prussia, where the first permanent bird observatory was established.

1901: *Bird Watching* by **Edmund Selous** (1857–1934).

1901: *How to Know the Indian Ducks* by **Frank Finn** (1868–1932).

1901: *The Home Life of Wild Birds: a New Method of the Study and Photography of Birds* by **Francis Hobart Herrick** (1858–1940).

1901: creation of the **Royal Australasian Ornithologists Union**, a society whose goal is the protection of birds of Autralia and surrounding regions.

1902: extinction of the Auckland Island Merganser (*Mergus australis*).

1904: the **Bird-Land Camera**, created by **Oliver Gregory Pike** (1877–1963), goes on sale, the first camera dedicated to animal photography.

1903: the first aeroplane takes flight

BIRD WATCHING

Bird Watching

Wild Life at Home

236

1905

1905: **Ottó Herman** (1835–1914) promotes his method of data analysis for the observation of birds.

1905: founding of the **National Audubon Society** that federated the many Audubon Societies in the United States.

1906: founding of the **Dansk Ornitologisk Forening** (or Ornithological Society of Denmark).

Adventures in Bird Land

Ottó Herman's book

1907: the Zoological Record lists 1,760 zoological articles

1907: launch of the journal ***British Birds***, an independent publication devoted to the study of birds, which remains in print today.

1909: the British Museum had by now acquired 500,000 bird skins.

1909: **Richard Bowdler Sharpe** (1847–1909) described 18,939 species and subspecies of bird.

1909: **Auguste Ménégaux** (1857–1937) and **Louis Denise** (1863–1914) founded the ***Revue française d'ornithologie***.

1909: *Umwelt und Innenwelt der Tiere* by **Jakob von Uexküll** (1864–1944).

Richard Bowdler Sharp

1910

1910

1910: the creation of
the National Museum
of Natural History
in Washington

1912: *A Manual of Palaearctic Birds* by **Henry Eeles Dresser** (1838–1915).

1912: creation of the **Ligue pour la protection des oiseaux (LPO)** in France.

1912: *Catalogue des oiseaux d'Europe pour servir de complément à l'Ornithologie européenne de Degland et Gerbe* by **Édouard Trouessart** (1842–1927)

The Professionalization of Science

1914-1918: First
World War

1914: death of the last Passenger Pigeon (*Ectopistes migratorius*) in captivity.

1914: **Sir Julian Huxley** (1887–1975) published his first works on avian ethology.

1915

1916: convention between Canada and the United States for the protection of migratory bird species.

1918: *Appetites and Aversions as Constituents of Instincts* by **Wallace Craig** (1876–1954).

1918: death of the last Carolina Parakeet (*Conuropsis carolinensis*), a victim of overhunting due to the use of its feathers for women's fashions.

1920

1920 — 1920: *Territory in Bird Life* by **Eliot Howard** (1873–1940).

1920: creation of the *Revue d'histoire naturelle appliquée: l'Oiseau* by **Jean Théodore Delacour** (1890–1985).

1921: **Erwin Stresemann** (1889–1972) takes up the direction of the museum of zoology in Berlin.

1922: formation of the **International Council for Bird Preservation** which federated multiple ornithological organizations, its objective was to protect birds and their habitats. It was renamed **BirdLife International** in 1993.

1927: *Animal Ecology* by Charles Sutherland Elton (1900-1991) contributed to the popularization of ecological studies

1923: **International Congress for the Protection of Nature**.

1924: *Birds of Oklahoma* by **Margaret Morse Nice** (1883–1974).

1924: *Die Vögel Mitteleuropas* by **Oskar Heinroth** (1871–1945) and de **Katharina Heinroth** (1897–1989).

1928: **Reginald Ernest Moreau** (1897–1970) began to study the birds of Africa.

1930 — 1928: creation of the journal ***Alauda*** then in 1933 the Société d'études ornithologiques.

1930: creation of the Service central de recherches sur la migration des oiseaux by **Édouard Bourdelle** (1876–1960).

1931: *Check-list of Birds of the World* by **James Lee Peters** (1890–1952).

1932: creation of the **British Trust for Ornithology (BTO)** an organization devoted to the study of birds in the field and relying heavily upon work by volunteers.

238

1936: foundation of the Institute for the Study of Animal Behaviour

1932: **Lord Lionel Walter Rothschild** (1868–1937) sold the largest part of his ornithological collections to the American Museum of Natural History in New York.

1934: *A Field Guide to the Birds* by **Roger Tory Peterson** (1908–1996), his first field guide.

1936: *Inventaire des oiseaux de France* by **Noël Mayaud** (1899–1989), **Henri Heim de Balsac** (1899–1973) and **Henri Jouard** (1896–1938).

1940 — 1940: publication of the first volume of *La Vie des oiseaux* by **Paul Géroudet** (1917–2006).

1939-1945: Second World War

1942: *Systematics and the Origin of Species* by **Ernst Mayr** (1904–2005).

1946: Ernst Mayr describes 28,500 species and subspecies

1944: *La Vie des colibris* by **Jacques Berlioz** (1891–1975).

1945: *The Wren* by **Edward Allworthy Armstrong** (1900–1978).

1948: founding of the **International Union for Conservation of Nature** (IUCN).

1950 — 1949: *Les Oiseaux dans la nature* by **Paul Barruel**.

The Environmental Crisis

1950 —

1950: the **Royal Society for the Protection of Birds** has 30,000 members.

1950-1951: *The Birds of Greenland* by **Finn Salomonsen** (1909-1983).

1953: *The Herring Gull's World* by **Nikolaas Tinbergen** (1907-1988).

Nikolaas
Tinbergen

1960 —

1960: *Atlas der Verbreitung palaearktischer Vögel* by **Erwin Stresemann** (1889-1972) and **Leonid Aleksandrovich Portenko** (1896-1972).

1962: the publication of *Silent Spring* by Rachel Carson

1961: *Bird-Song. The Biology of Vocal Communication and Expression in Birds* by **William Homan Thorpe** (1902-1986).

1963: *On Aggression* by **Konrad Lorenz** (1903-1989).

1967: shipwreck of the Torrey Canyon

1965: *Avant que nature meure* by **Jean Dorst** (1924-2001).

Konrad
Lorenz

1970 —

1973: **Nikolaas Tinbergen** (1907-1988), Karl von Frisch (1886-1982) and **Konrad Lorenz** (1903-1989) shared the Nobel prize for physiology or medicine for their work on animal behaviour.

1976: **Michel Brosselin** initiated the creation of the **Union nationale des associations ornithologiques**.

239

TIMELINE

1980 —

1984: *Atlas of Australian Birds* by the Royal Australasian Ornithologists Union (updated in 2002).

1990: the *Zoological Record* lists 14,000 ornithological articles

1990 —

1990: *Phylogeny and Classification of Birds: a Study in Molecular Evolution* by **Charles Gald Sibley** (1917-1998) and **Jon Edward Ahlquist**.

1990: *Distribution and Taxonomy of Birds of the World* by **Charles Gald Sibley** (1917-1998) and **Burt Monroe** (1930-1994).

1994: the LPO begins publishing **Ornithos.**

1997: the first human cases of avian flu (H5N1)

2000 —

2002: the RSPB has more than one million members.